T0406586

Insecticides–Soil Microbiota Interactions

Naga Raju Maddela • Kadiyala Venkateswarlu

Insecticides–Soil Microbiota Interactions

 Springer

Naga Raju Maddela
Sun Yat-sen University
East Campus, School of Environment
  Science & Engineering
Guangzhou, China

Kadiyala Venkateswarlu
Sri Krishnadevaraya University
Anantapur, AP, India

ISBN 978-3-319-66588-7   ISBN 978-3-319-66589-4   (eBook)
DOI 10.1007/978-3-319-66589-4

Library of Congress Control Number: 2017950497

Printed on acid-free paper

This Springer imprint is published by Springer Nature
The registered company is Springer International Publishing AG
The registered company address is: Gewerbestrasse 11, 6330 Cham, Switzerland

Preface

The excess use of fertilizers, pesticides, other agrochemicals and animal manure greatly contribute to significant changes in soil and contamination of the agro eco-system as well as the neighboring ecosystems. The estimated worldwide annual sales of pesticides rapidly increased from 1.0 to 35 billion US$ between 1960 and 2000. In India alone, the total consumption of pesticides increased from ~70,000 to ~80,000 tonnes (active ingredient) during 2005–2010. Since agrochemicals are sprayed or spread across entire agricultural fields, 95–98% of these applied chemicals will reach soil which is the ultimate sink for all the sources. Pesticides are highly toxic, and they contaminate waters and soils, consequently limiting microorganisms that are largely implicated in soil fertility besides harming several beneficial insects. The situation would be too worse when it rains as water interacts with these toxic agrochemicals which then leach into ground water sources or get washed through surface runoff into nearby water bodies. Subsequently, humans and other animals are also affected when they consume such polluted waters. During 1961–2002 alone, the worldwide consumption of fertilizers increased from 30 to 140 million tones. Although fertilizers and animal manure are not directly toxic, they usually contain excess chemical nutrients in the form of nitrogen and phosphorus. These excess nutrients also affect the water quality, resulting in depletion of dissolved oxygen in water and consequent development of eutrophication followed by death of fishes and other aquatic animals.

Even though less crop land (2.4%) is planted with cotton in the world, monocrop-ping of cotton accounts for 24% of the global use of pesticides. Two major insecti-cides, acephate and buprofezin, are widely used to control several important insect pests of cotton in surrounding regions of Nandyal in Kurnool district of Andhra Pradesh, India. Acephate is a foliar organophosphorus (OP) insecticide, used primar-ily for the control of aphids and also leaf miners, caterpillars, sawflies and thrips. Buprofezin is a thiadiazole insecticide that controls mealybugs, leafhoppers and whitefly. In some situations, the above two insecticides are applied repeatedly and in excess doses by farmers to control the insect pests. Over time, repeated and indis-criminate applications of pesticides not only increase pest resistance, but also exert several deleterious effects on soil ecosystem. Assessing such nontarget effects of acephate and buprofezin in soil is the main focus of the present book. Activities of soil enzymes such as cellulases, amylases, invertase, proteases and phosphatases were chosen as indicators for investigating the extent of soil pollution of acephate and buprofezin. Also, we provided preliminary data on utilization of the selected insecticides by a bacterial strain isolated from soil following enrichment. Thus, the main objective of this project is to provide detailed information on insecticides–soil microbiota interactions. We believe that the present information is very useful for the researchers as well as agriculturists in understanding the effects caused by extensive and intensive use of the two selected insecticides in agricultural fields.

For the convenience of the readers, information related to the above aspects has been divided into ten chapters in this book. It is expected that the present book serves as a single source of information for those who look for information on interactions between insecticides and soil microorganisms. Chapter 1 provides an overview and background in selection of the present topic. This chapter contains a brief discussion on cotton cultivation in the selected region and global usage of pesticides and fertilizers. Chapter 2 outlines an account as to how intensive agricultural practices pollute soils and cause detrimental changes in soil fertility and health. It considers soil enzymes as indicators of soil pollution. Chapter 3 provides details on soil samples, insecticides, fertilizers and soil enzymes selected for this study. This chapter also deals with information about the experimental setup used for studying the nontarget effects of selected insecticides toward soil enzyme activities. Chapters 4, 5, 6, 7, 8, and 9 discuss the impact of acephate and buprofezin on cellulases, amylases, invertase, proteases, urease, and acid and alkaline phosphatases, in that order. Each chapter provides the protocol for assay of an enzyme in soil, discusses nontarget effects of the insecticides on enzyme activities after single, two or three applications, and interaction effects of insecticide combinations and nutrient amendments as well. Chapter 10 deals with the impact of soil microorganisms on persistence of insecticides. It outlines the utilization of the selected insecticides by a bacterium, *Pseudomonas* sp., isolated from soil following selective enrichment.

Guagnzhou, China   Dr. Naga Raju Maddela, PhD
Nellore, AP, India   Prof. Kadiyala Venkateswarlu, PhD

Contents

Chapter 1
Introduction

Cotton (*Gossypium hirsutum* L.), an important fiber-yielding crop of global economic importance, is extensively grown in tropical and subtropical regions of more than 80 countries in the world. As it provides fiber for cloth, edible oil from the seed, and protein-rich seed cake, cotton can be described as a three-in-one wonder crop. Cotton is the major crop occupying 8.5 million ha in India, and on an average this crop is cultivated in 25,000 ha in Kurnool district of Andhra Pradesh alone. It is also one of the chief commercial crops grown, almost as a monocrop, in Nandyal division of Kurnool district. In fact, cotton is the fourth crop next to groundnut, Bengal gram and jowar cultivated in this area.

The vast economic damage caused to cotton by many insect pests such as aphids, bollworms, stem borers, thrips, planthoppers, green leafhoppers, mealy bugs, jassids, and silver leaf white flies is a serious concern. Several insecticides are used on need basis for the effective control of these insect pests of cotton. In particular, two insecticides, acephate (Hythene®), an OP compound, and buprofezin (Applaud®), a thiadiazine, are widely used in the recent years to combat major insect pests of cotton. Also, acephate and buprofezin are applied repeatedly and, sometimes, in combination at a dose of 0.292–0.584 and 0.25 kg ha^{-1}, respectively. In particular, acephate, a nonphytotoxic foliar spray OP insecticide, has been used for controlling a wide range of biting and sucking insects, especially aphids, leaf miners, lepidpterous larvae, saw flies and thrips. Also, it is more effective in controlling jassids and bollworms on cotton, with a residual systemic activity of about 10–15 days at the recommended rates. Furthermore, acephate with its moderate persistence dissipates rapidly. On the other hand, buprofezin is extensively used on cotton besides rice, citrus, mango, vegetables, ornamentals, grape and tea crops for controlling a variety of insect pests such as scale, mealy bug, jassids (leafhoppers), plant hoppers, silver leaf white flies at a dose of 250 g ha^{-1}.

The external agricultural inputs such as mineral fertilizers, organic amendments, microbial inoculants, and pesticides aim at the ultimate goal of maximizing productivity and economic returns. In particular, fertilizer use on cotton crop accounts for

N.R. Maddela, K. Venkateswarlu, *Insecticides–Soil Microbiota Interactions*,
DOI 10.1007/978-3-319-66589-4_1

6.0% (1.01 million tons) of total fertilizer consumption. The use of fertilizers on irrigated cotton is higher than on rainfed cotton. Thus, the share of cotton in total fertilizer consumption under irrigated and rain-fed conditions is ~3.3% and ~2.7%, respectively. Although mineral fertilizers have limited direct effects, their application enhances soil biological activity through increases in system productivity, crop residue return, and soil organic matter. Another important indirect effect, especially of nitrogen (N) fertilization, is soil acidification which has considerable negative effects on soil organisms.

Consumption of agrochemicals in the world is thus increasing on regular basis. For examples, the global use of pesticides and fertilizers at least during 2004–2014 increased dramatically by 8.4% and 21.9%, respectively (Fig. 1.1). Especially, the

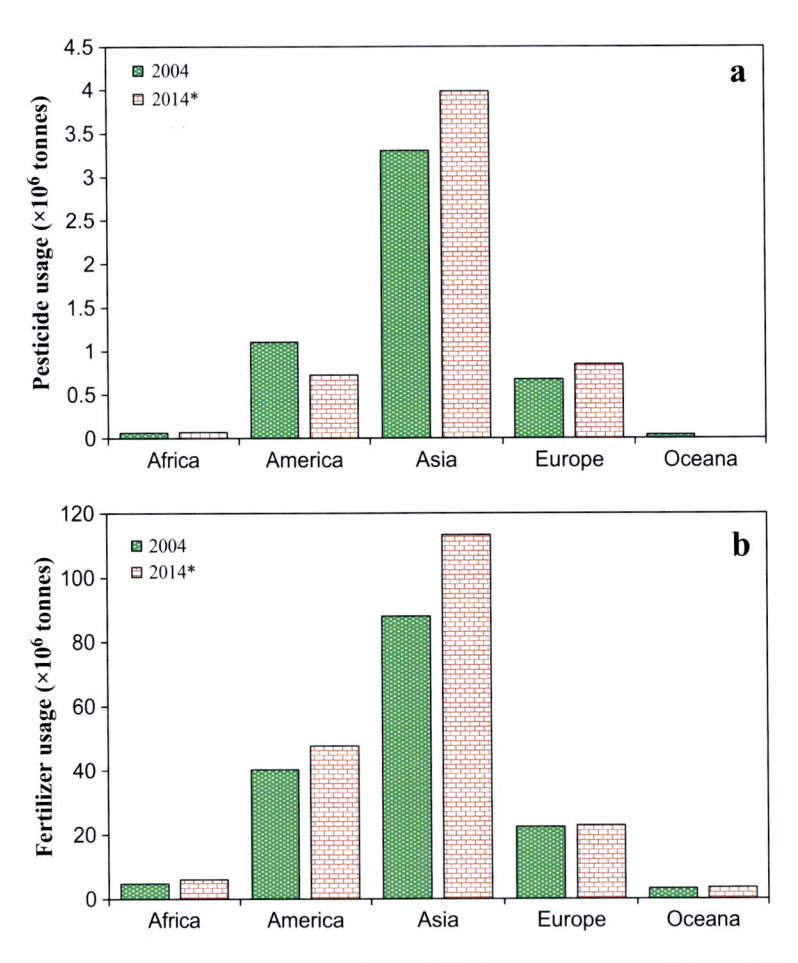

Fig. 1.1 Consumption of (**a**) fertilizers and (**b**) pesticides in different regions of the world for the years 2004 and 2014 (FAO 2016). *Excludes data from USA, Australia and New Zealand

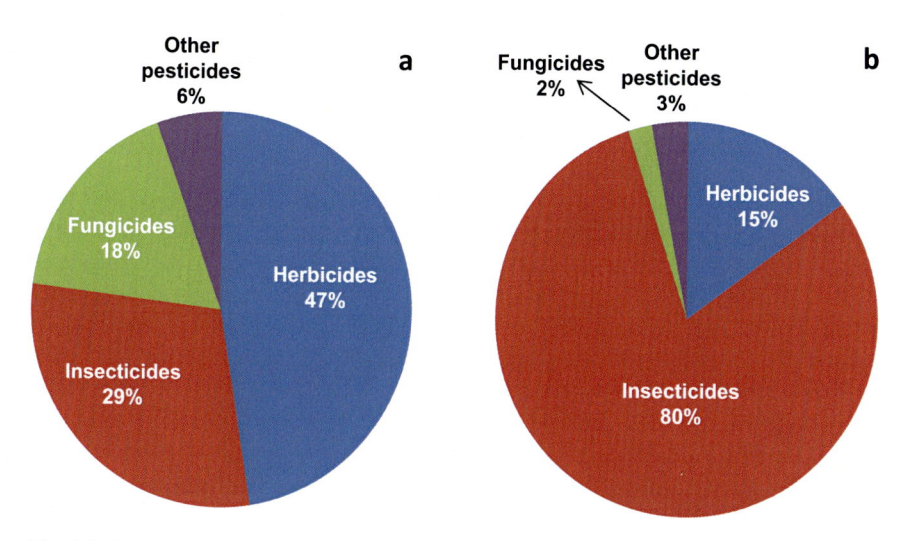

Fig. 1.2 Consumption of pesticides in (**a**) the world and (**b**) India (De et al. 2014)

extent of both pesticide and fertilizer consumption in Asia was 55–70% of their total use in the world (Fig. 1.1). Majority of pesticides used in India are insecticides (80%) followed by herbicides (15%) (Fig. 1.2b), a scenario that is different from world's consumption (Fig 1.2a). It has been very well established that about 99–99.9% of the pesticides applied even as a foliar spray onto an agricultural crop is certain to reach the soil. Thus, soil serves as the ultimate sink for all these agro-chemicals applied in the modern agriculture.

The occurrence of xenobiotics, pesticides in particular, in the soil environment will have far-reaching consequences on total population of microorganisms and their biological activities mainly implicated in soil fertility. A few significant effects of herbicides on soil organisms have been documented, whereas negative effects of insecticides and fungicides are more common. Generally speaking, application of pesticides at recommended rates have little or no effect on soil fertility and health. In contrast, when pesticides are applied to soil at rates higher than the recommended doses over long periods, significant effects on soil health can be expected. OP insec-ticides, in general, have a range of effects including changes in bacterial and fungal populations in soil; their application therefore warrants strict regulation. It has become increasingly possible to isolate microorganisms that are capable of degrad-ing xenobiotics and recalcitrant compounds from environments polluted with these toxic chemicals. The nutritional requirements that must be considered for microbial growth are the sources of carbon, energy and nitrogen, other major mineral nutrients such as sulfur, phosphate, potassium, magnesium, calcium, and trace metals. Isolation of indigenous bacteria capable of metabolizing OP compounds has there-fore received considerable attention because OP-degrading bacteria provide an environmentally friendly approach of in-situ detoxification. The standard method employed since the earliest days of microbiology using selective cultures has its drawbacks; the most critical of which is that many bacteria are capable of metabolizing substrates but cannot utilize such substrates as sole sources of energy or of carbon for proliferation.

Despite the fact that the two insecticides, acephate and buprofezin, are used, both extensively and intensively, on a major commercial agricultural crop like cotton, there are no reports on their nontarget effects and microbial degradation of these insecticides. For assessing the impact of insecticide usage in agriculture, measures of microbial activities are generally considered as good indicators of the degree of soil pollution. The most commonly used methods for measuring microbial activities include soil respiration rates, enzymatic activities related to the microbial cycling of nutrients, and certain biogeochemical functions. Soil enzyme activities, in particular, are useful integrative indicators of soil health and have been used widely to assess the effects of management practices on soil biological functioning. Indeed, microbial activities refer to all the metabolic reactions and interactions among the microflora and microfauna that exist in soils. Enzymes in soil are either intracellular and present as a component of viable soil organisms, or extracellular and bound to clay or humic acids. Enzyme activity in soil is regulated by pH and microbial biomass, and is correlated to soil organic matter, soil moisture content, and soil compaction. Analysis of the activity of soil enzymes thus provides information on biochemical processes occurring in soil.

Soil enzyme activity is variable in time and limited by available substrate supply, and provides useful linkage between composition of microbial community and carbon processing. About 90% of energy in the soil environment flows through microorganisms. Thus, information on soil enzyme activities, used to determine soil microbiological functioning, is very important for determination of soil quality and health since abundance and activities of microorganisms reveal the degree of soil fertility. The status of an enzyme in soil may determine how pesticides affect its activity. Available evidences indicate that both bacterial growth and the microbiological activities are negatively influenced by high concentrations of pesticides whereas lower concentrations stimulated the overall metabolic soil activity. In this direction, several studies were conducted with insecticides to evaluate their effects on soil populations of bacteria, fungi and actinomycetes, and soil enzymes after single or repeated applications of several insecticides. However, virtually no information is available in the literature on acephate and buprofezin, and their nontarget effects of acephate and buprofezin towards microbial activities in soil or microbial degradation of these two widely used insecticides. In the following chapters, we present recent findings on insecticide–soil microbiota interactions with emphasis on the impact of acephate and buprofezin on activities of soil enzymes such as cellulases, amylases, invertase, proteases, urease and phosphatases, and utilization of these insecticides by a bacterium isolated from soil following enrichment culture method.

References

De A, Bose R, Kumar A, Mozumdar S (2014) Targeted delivery of pesticides using biodegradable polymeric nanoparticles. SpringerBriefs in Molecular Science. doi:10.1007/978-81-322-1689-6, ISBN: 9788132216896 (online). Springer, p. 99

FAO STAT Pesticide use (2016) Retrieved from: http://www.fao.org/faostat/en/#data/RP (Online resource). Accessed April 2017

Chapter 2
Soil Enzymes: Indicators of Soil Pollution

Soil Fertility and Health

Soil health, also referred to as soil quality, is a status of soil conditions which functions appropriate to its environment. Soil health testing is an assessment of this status. Soil quality largely depends on soil biodiversity, and it can be improved via soil management practices. Generally speaking, soil is a dynamic system that appears to be in a biological equilibrium. However, this equilibrium is precarious since any disturbance in soil ecosystem has the potential to change the microbial populations and activities. In reality, soil microorganisms have primary role in the environment through degradation of plant and animal residues. They play a pivotal role in many soil biological processes, including nitrogen transformation, organic matter decomposition, and nutrient release and their availability (Sparling 1985; Lee and Pankhurst 1992). Thus, activities of microorganisms in soil are essential to global cycling of nutrients, which has a direct bearing on soil fertility and health. Since majority of biochemical transformations in soil result from microbial activity, any compound that alters the number or an activity of microbes could affect soil biochemical process and ultimately soil fertility and plant growth (Cohen et al. 1984).

It has long been known that the abundance and activities of microorganisms reveal the degree of soil development (Powlson et al. 1987). About 90% of energy in the soil environment flows through microorganisms (Heal and Lean 1975). More broadly, the term "soil microbial activity" implies to the overall metabolic activity of all microorganisms inhabiting soil, including bacteria, fungi, protozoa, algae and microfauna (Nannipieri et al. 1990). But, bacteria are the most predominant organisms in soil. Healthy and fertile soil may contain 1.0×10^9 bacteria g^{-1} soil. In a recent study, Vrieze Jop de (2015) observed more than 33,000 bacterial and archaeal species on sugar beet roots. Nonetheless, healthy soil gives us clean air and water, bountiful crops and forests, productive grazing lands, diverse wildlife, and beautiful landscapes. Thus, to maintain soil health, it is important to avoid contamination of soils by anthropogenic chemicals that have deleterious impact on soil organisms and their processes.

© Springer International Publishing AG 2018
N.R. Maddela, K. Venkateswarlu, *Insecticides–Soil Microbiota Interactions*,
DOI 10.1007/978-3-319-66589-4_2

Enzymes in soils are known to play a substantial role in maintaining soil health and its environment. Several enzymes, known to be present in soil, catalyze organic matter turnover. These enzymes are produced by various organisms, and are mainly of bacterial and fungal origin, and act intra- or extra-cellularly. Only a small fraction is derived from plants and/or animals. Physical status of enzymes in soil is in different forms. For example, enzymes that are stabilized in soil matrix accumulate or form complexes with organic (humus), inorganic (clay) or organic-inorganic complexes. Also, in a given soil, enzyme activity does not necessarily correlate with the microbial biomass or respiration, since 40–60% of enzyme activity can come from stabilized enzymes. Enzyme activity in soil is regulated by pH and microbial biomass (Dick et al. 1988), which is correlated to soil organic matter and soil moisture content (Harrison 1983) as well as soil compaction (Karaca et al. 2000). However, soil enzyme activity is variable in time and limited by available substrate supply (Tateno 1988; Degens 1998), and may provide useful linkage between microbial community composition and carbon processing (Waldrop et al. 2000).

Soil enzymes also play a crucial role in degrading litter and "foreign" substances. Thus, the role of enzymes, in terms of soil ecosystem, is increasingly important and is defined by the relationships between soil enzymes and the environmental factors that affect their activities (Paul and Mclaren 1975; Burns 1982). The enzymes most often present in soil are cellulases, dehydrogenase, phosphatases, amylases, catalase, xylanase, pectinase, saccharase, proteases and urease. Although most of soil enzyme research evolved without any consideration to the ecological implications, soil enzymes are highly useful in describing and making predictions about an ecosystem's function, quality and the interactions among subsystems. Thus, information on soil enzyme activities used to determine soil microbiological characteristics is very important for soil quality and healthy.

Intensive Agriculture and Soil Pollution

'Soil pollution' is defined as 'the presence of toxic chemicals (pollutants or contaminants) or waste materials in soil, in high enough concentrations that pose a risk to human health and/or the ecosystem' (Ramakrishnan et al. 2010). It is typically caused by improper disposal of wastes. The most common form of chemicals involved includes petroleum hydrocarbons, polycyclic aromatic hydrocarbons, solvents, pesticides and heavy metals. All such pollutants are coming from various activities such as agrochemicals (pesticides and fertilizers) applied during intensive and extensive farming, oil drilling, mining, accidental oil spills, corrosion of underground storage tanks, acid rains, industrial accidents, road debris, and waste disposal (nuclear wastes, coal ash, ammunitions and agents of war, etc.). It is a well known fact that soil pollution is mostly correlated with the degree of industrialization and intense use of chemicals. The two important concerns of soil pollution are health risks from direct contact, and secondary contamination of water supplies within and underlying soil. Nevertheless, soil pollution affects plants, animals and humans alike. Even though several kinds of pollutants are causing damage to the ecosystem, agricultural pollutants are fatally flawed because of declining crop yields and massive environmental impacts. Agrochemicals not only contaminate or degrade the environment and surrounding ecosystems, but also cause injury to humans and their economic interests. Management practices play vital role in the amount and impact of agricultural pollutants. The two important agro-pollutants in global agricultural practices are therefore pesticides and fertilizers.

It is well documented that the agricultural chemicals especially pesticides are perhaps the largest group of poisonous substances being disseminated throughout our environment (El-Aswad et al. 2001). Indiscriminate methods of insecticide applications are rampant, often allowing high loads of xenobiotics to reach the soil matrix. Pesticides may enter the soil either directly or indirectly (Burns 1975), and it is apparent that crop protection results in accumulation of these residues in the soil which is ultimately the sink for all the anthropogenic compounds. The primary objective of modern agriculture is to produce a reliable supply of wholesome food to feed the burgeoning world population, safely and without adverse effects on the environment. Thus, the intensified agriculture in developing countries has, therefore, dictated the increasing use of agrochemicals to meet growing food demands.

Pesticides, insecticides in particular, enable to achieve higher crop yields by controlling major insect pests. Numerous reports have indicated that at the present time, the most widely used pesticides belong to the OP group. These compounds are among the most common chemical classes used in crop and livestock protection and account for an estimated one-third of worldwide insecticide scales (Singh and Walker 2006). Historically, to protect crops, livestock, and human health approximately 150 different OP chemicals were used at least until the year 2004 (Casida and Quistad 2004). In agriculture, pesticides are applied directly to crops or during post-harvest storage or transport. In addition, soil may be treated pre-planting and during plant growth for control of pests, which pose threat to the major agricultural crops.

In recent intensive agriculture, pesticides are applied in many ways of delivery to the target, with concentrations and formulation varying to suit the mode of application and the target. Indeed, only 1% of the applied pesticides hit the target. Most of the pesticides applied to crops are deposited, not on the pest, but on the surrounding soil, particularly the 2–5 cm of top soil with concomitant side effects on nontarget microorganisms. In the global agricultural system, several insecticides can be used on need basis. Also, pesticides are often applied several times during a single crop season. Thus, the use of pesticides has become an integral and economically essential part of modern agriculture. The same insecticide should not be repeatedly sprayed to prevent the insect developing resistance to insecticides. Studies in the past have mostly focused on fate, metabolism and impact on soil following single applications of pesticides for short periods. As pesticides are designed to be biologically active, their continuous use might affect soil microflora either by changing their properties or the number which may lead to impairment of soil fertility. While useful data have been generated, investigations reflecting real situations, involving heavy repeated long-term applications, have been scanty. Since soil is the most important agricultural resource, next to water, it is important to study the possible effects of specific practices on soil properties.

A detrimental biological activity of soil appears to be the simultaneous application of high levels of mineral fertilizers with chemicals for pest/weed control. Generally, a fertilizer is either from natural or synthetic origin, and is applied to soils or to plant tissues (mostly leaves) to supply nutrients essential for plant growth. From agricultural sources point of view, fertilizers and animal manure are the primary sources of nutrient pollution. Both are rich in nitrogen and phosphorous. The effects of nutrient pollution are much severe with rains. For instance, water and soil containing nitrogen and phosphorous wash into nearby surface waters or leach into ground waters. In addition, both fertilized soils and livestock can be significant sources of ammonia and nitrogen oxides. In these two nitrogen-based and gaseous forms of pollutants, ammonia can be very harmful to aquatic life if large amounts are deposited in surface waters, whereas nitrous oxide is a potent greenhouse gas. Generally, upon the application of nitrogen-based fertilizer to soil, very little fraction is converted to the agricultural produce and other plant matter, and remainder accumulates in soil or lost as runoff.

When nitrogen-containing fertilizers are applied at high rates, they tend to combine with highly water-soluble nitrates, and leads to increased runoff. If the ground water contains above 10 mg L^{-1} nitrates, it causes "blue baby syndrome" which is also called as "acquired methemoglobinemia". On the other hand, complex fertilizers usually contain ammonium nitrate, phosphorous as P_2O_5 and potassium as K_2O. Certain heavy metals such as arsenic (As), lead (Pb) and cadmium (Cd) that are present in trace amounts in rock phosphate mineral get transferred to super phosphate fertilizer. Due to excess use of phosphate fertilizers and insoluble nature of heavy metals, these heavy metals accumulate in soil above their toxic levels and become an indestructible poison for crops. It has also been found that excess and indiscriminate use of complex N-P-K fertilizers reduce the quantity of vegetables and crops grown in soil over the years (Soil Pollution 2017). Nutritional status of plant products is also greatly affected by indiscriminate use of N-P-K fertilizers. For instance, protein content and carbohydrate quality of wheat, maize, grams decrease with chemical fertilization (Soil Pollution 2017). Excess content of potassium in soil decreases contents of vitamin C and carotene in certain commercial crops such as vegetables and fruits. Moreover, vegetables and fruits cultivated in over-fertilized soils are highly susceptible to diseases caused by insects. In all, use of fertilizers beyond the recommended doses should be regulated strictly to reduce their deleterious effects on soil ecosystem, particularly soil fertility and crop yield.

Excessive use of natural resources and large-scale synthesis of xenobiotic compounds have generated a number of environmental problems such as contamination of air, water and terrestrial ecosystems, harmful effects on different biota, and disruption of biogeochemical cycling. With regard to the pesticides, the heavy usage may adversely affect the natural environment, including soil ecosystem, disturbing its homeostasis. Precisely, they affect the soil properties, which in turn may affect the crop yield in due course. In general, harmful effects are generally related to the concentrations of pesticide applied, and correct dosage is often the key to their successful use without hazardous side effects. For instance, treatment of soil with 50 and 100 µg g^{-1} of methyl parathion showed a clear initial reduction in bacterial population count, but there was a gradual increase in bacterial count on the 35th day of incubation in all the three concentrations (Bindhya et al. 2009).

Singh et al. (2002) reported that the measured soil microbial parameters (enzyme activities and total microbial biomass) were stable in pesticide-free control soils throughout the 90-day incubation period, but they were all adversely affected by the presence of chlorothalonil in soil. However, the negative effects from fenamiphos or chlorpyrifos on soil microbial characteristics were either very small or insignificant. Michael and Turgeona (1978) suggested that the rates of glucose utilization, nitrification of ammonium, amylase synthesis were significantly lower in soil underlying treated turf than in control soil. Surprisingly, although short-lived inhibitory effects on activities of microbes and enzymes were caused by the insecticides, the soil indigenous microbes can tolerate the chemicals used for control of soil pests (Tu 1995). Cernakova et al. (1992) found that both bacterial growth and the activities under study were negatively influenced by high concentrations of actellic whereas lower concentrations stimulated the overall metabolic soil activity. Similarly, the electron transport system/dehydrogenase activity displayed a negative correlation with triazophos, bensulfuron-methyl, chlobenthiazone concentration of pesticide increased (Xie et al. 2004). Thus, the application of agrochemicals, in excess of recommended doses, to increase crop productivity may cause a variety of negative environmental changes including significant inhibition of soil biological activity (Wyszkowska and Kucharski 2004).

An intriguing discovery is that the impact of long-term DDT pollution in soil by using different criteria (Megharaj et al. 2000). The criteria used included chemical analysis of DDT residues, microbial biomass, and dehydrogenase activity, viable counts of bacteria and fungi, and density and diversity of algae. The experimental results indicated that the viable counts of microalgae and bacteria decreased with increasing DDT contamination, while fungi, microbial biomass, and dehydrogenase activity increased in soil contaminated at medium levels (27 mg DDT residues kg^{-1} soil). All of the tested parameters were greatly inhibited in the high-level contaminated soil (34 mg DDT kg^{-1} soil). More recently, the effect of fenamiphos, a widely used OP pesticide, on important soil microbial activities such as dehydrogenase, urease and potential nitrification in four soils from Australia and Ecuador was studied by Caceres et al. (2009). The results showed that fenamiphos in general was not toxic to dehydrogenase as well as urease up to100 mg kg^{-1} soil. However, potential nitrification was found to be highly sensitive to fenamiphos with a significant inhibition recorded even at 10 mg kg^{-1}. In general, nitrification activity in soils was decreased with an increase in fenamiphos concentration. In contrast, dimethoate and malathion when added to soil at 10 and 100 µg g^{-1} caused an initial stimulation of CO_2 production, and there was an increase in total counts of bacterial propagules.

Application of all insecticides increased bacteria producing phosphatases from the first week until week 4 after the application; bacteria then returned to the original levels (Congregado et al. 1979). There is an evidence that the repeated applications of some herbicides like atrazine, 2,4-D, paraquat, and trifluralin over many years may compound a negative impact, change microbial community structure, or build-up biodegradation capacity (Pankhurst 2006). Application of these agrochemicals in pest control can lead to serious pollution of soil and water environments as these chemicals and their residues alter the environment associated with microorganisms. As such, there is no information available on the interaction between the selected insecticides, viz., acephate and buprofezin, and microorganisms that are implicated in the transformation of carbon, nitrogen and phosphorous in soil.

Soil Enzymes as Indicators of Pesticide Pollution

An examination of our environment associated with microorganisms, largely implicated in soil fertility, and their interactions with different pollutants provides insights into the effects of these chemicals that alter growth and development of organisms, and the response of living organisms (Ramakrishnan et al. 2010, 2011). The interactions of different pollutants with microorganisms may vary in nature and magnitude. Investigations to assess the effects of environmental pollutants include studies on nontarget influence of toxicants toward ecologically beneficial microorganisms, and those that deal with the impact of soil microorganism on the persistence of the toxic chemicals. In this direction, there is a considerable interest in the study of enzyme activities of soils (Burns 1978; Greaves and Malkones 1980) because soil enzyme activities are useful integrative indicators of soil health, and have been used widely to assess the effects of management practices on soil biological functioning (Dick 1997).

The role of microbial activity in the development and functioning of soil ecosystem is, therefore, inevitable, and changes in soil microbial activity may be an indication of changes in soil health (Pankhurst et al. 1995). The essential point to remember about enzymes is that they are frequently referred to as markers of soil environment purity (Aon and Colaneri 2001), since such activities may reflect the potential capacity of a soil to perform certain biological transformations important to soil fertility. As has already been indicated, enzymes participate in numerous biochemical processes occurring in soil, and as a result they are sensitive to all environmental changes caused by natural and anthropogenic activities (Trasar-Capeda et al. 2000). For example, pesticides are often applied several times during one crop season, and a major part of them always reaches soil and eventually affects soil enzyme activities.

In modern agriculture, it has become a common trend to apply different groups of pesticides, either simultaneously or in succession, for effective control of a variety of pests (Venkateswarlu 1993). It is now known that monitoring of the pedosphere using the methods based on enzymatic tests enables a complex assessment of changes in the soil environment under the influence of anthropogenic factors (Taylor et al. 2002). Hence, soil enzyme activities are useful integrative indicators of soil health and have been used widely to assess the effects of management practices on soil biological functioning (Dick 1997). Enzymes respond to soil management practices long before other soil quality indicator changes are detectable. For some years now, measurements of microbial biomass and various enzyme systems have been widely used to diagnose the soil state and to describe the effect of different influences of pollutants, agricultural management, and land use. It is also worth considering that enzymatic activities as caused by soil microbial activities are sensitive indicators to detect changes occurring in soils (Gonzalez et al. 2007).

Possible indicator value of the microbial parameters for environmental stress, in general, was investigated through microbial processes including soil enzyme activities (Bandick and Dick 1999; Tscherko and Kandeler 1999). Perhaps the most valuable single use of soil enzymes has been to assess the effects of various inputs on the relative "health" of soil. Numerous studies have been conducted to determine changes in activities of a soil enzyme caused by acid rain, heavy metals, pesticides, and other industrial and agricultural chemicals. Nonetheless, soil enzyme activities commonly correlate with microbial parameters (Frankenberger and Dick 1983) and have been shown to be sensitive indices of long-term pesticide effects. In this direction, various reports have indicated that under field and/or laboratory conditions, insecticides applied at commercially recommended rates, exerted an adverse effect on microbiological properties of soil as manifested by the observed altered enzymatic activities.

A perusal of the literature reveals the effects of insecticides on soil enzyme activities, an important parameter which helps in maintaining soil health and fertility. Dilly and Munch (1998) opined that the biomass-specific respiration and metabolic quotient which combines microbial activity and population is a more sensitive indication of soil pollution. Some evidences have suggested the use of exogenous microorganisms or enzymes as bioassay activity measurement needs careful consideration of the season and prevailing weather conditions (Ronnpagel et al. 1998; Brohon et al. 2001), because climatic factors often determine in situ variation of soil microbial activities (Insam 1990). Status of an enzyme in soil may determine how pesticides affect its activity. The sum of all the chemical reactions that occur within all living organisms is termed 'metabolism', the major new consequence of these reactions in microorganisms is the synthesis of a new cell.

To summarize, metabolic reactions that release energy, in the form of ATP, from the breakdown or degradation of a substrate (e.g., complex organic molecules) are catabolic reactions, while the ones that use energy to assemble smaller molecules and produce biosynthetic building blocks are called anabolic reactions; and when there is production of ATP and precursors of biosynthetic building blocks are called amphibolic reactions. All these bio-chemical reactions that occur both outside and inside the cell are precisely controlled by some governing factors such as the enzymes. More importantly, an enzyme is a biological catalyst, a substance that accelerates the rate of a specific chemical reaction. Application of pesticides at recommended rates have little or no effect on soil enzyme activities (Ladd 1985; Schäffer 1993; Nannipieri 1994; Dick 1997). In contrast, activities of enzymes were shown to be significantly affected when pesticides are applied to a soil at levels higher than the recommended rates over long periods (Voets et al. 1974; Rai 1992; Sannino and Gianfreda 2001; Megharaj 2002).

References

Aon MA, Colaneri AC (2001) Temporal and spatial evolution of enzymatic activities and physico-chemical properties in an agricultural soil. Appl Soil Ecol 18:255–270

Bandick AK, Dick RP (1999) Field management effects on soil enzyme activities. Soil Biol Biochem 31:1471–1479

Bindhya R, Sunny SA, Thanga VSG (2009) In vitro study on the influence of methyl parathion on soil bacterial activity. J Environ Biol 30:417–419

Brohon B, Delolme C, Gourdon R (2001) Complementarity of bioassays and microbial activity measurements for the evaluation of hydrocarbon-contaminated soils quality. Soil Biol Biochem 33:883–891

Burns RG (1975) Factors affecting pesticide loss from soil. In: Paul EA, Mclaren AD (eds) Soil biochemistry, vol 4. Marcel Dekker, Inc, New York, pp 103–141

Burns RG (1978) Soil enzymes. Academic, New York, p 370

Burns RG (1982) Enzyme activity in soil: location and a possible role in microbial ecology. Soil Biol Biochem 14:423–427

Caceres T, He W, Megharaj M, Naidu R (2009) Effect of insecticide fenamiphos on soil microbial activities in Australian and Ecuadorean soils. J Environ Sci Health B44:13–17

Casida J, Quistad G (2004) Organophosphate toxicology: safety aspects of non-acetyl cholinesterase secondary targets. Chem Res Toxicol 17:983–998

Cernakova M, Kurucova M, Fuchsova D (1992) Effect of the insecticide actellic on soil microorganisms and their activity. Folia Microbiol 37:219–222

Cohen SZ, Creeger SM, Carsel RF, Enfield CG (1984) Potential pesticide contamination of ground water from agricultural uses. In: Kruger RF, Seiber JN (eds) Treatment and disposal of pesticide waste. American Chemical Society, Washington, DC, pp 297–325

Congregado F, Simon-Pujol D, Juarez A (1979) Effect of two organophosphorus insecticides on the phosphate dissolving soil bacteria. Appl Environ Microbiol 37:169–171

Degens BP (1998) Microbial functional diversity can be influenced by the addition of the simple organic substances to soil. Soil Biol Biochem 30:1981–1988

Dick RP (1997) Soil enzyme activities as integrative indicators of soil health. In: Pankhurst CE, Doube BM, Gupta VVSR (eds) Biological indicators of soil health. CAB International, Wallingford, pp 121–156

Dick RP, Rasmussen PE, Kerle EA (1988) Influence of long-term residue management on soil enzyme activities in relation to soil chemical properties of wheat-follow system. Biol Fertil Soils 6:159–164

Dilly O, Munch JC (1998) Ratios between estimates of microbial biomass content and microbial activity in soils. Biol Fertil Soils 27:374–379

El-Aswad AF, Attia AM, Khalil AI (2001) Influence of malathion and metribuzin on microbial populations and their cellulolytic activities during the composition of vegetable residues. Alexandria J Agric Res 46:253–268

Frankenberger WT Jr, Dick WA (1983) Relationships between enzyme activities and microbial growth and activity indices in soil. Soil Sci Soc Am J 47:945–951

Gonzalez MG, Gallardo JF, Gomez E, Masciandaro G, Ceccanti B, Pajares S (2007) Potential universal applicability of soil bioindicators: evaluation in three temperate ecosystems. CI Suelo (Argentina) 25:151–158

Greaves MP, Malkones HP (1980) Effect on soils microflora. In: Hance RJ (ed) Interaction between herbicides and the soil. Academic, London, pp 223–253

Harrison AF (1983) Relationship between intensity of phosphatase activity and physico-chemical properties in woodland soils. Soil Biol Biochem 15:93–99

Heal OW, Lean SFM Jr (1975) Comparative productivity in ecosystems – secondary productive. In: Van Dobbeu WH, Lowe-Mc Connell RH (eds) Unifying concept in ecology. Junk, The Hague, pp 89–108

Insam H (1990) Are soil microbial biomass and basal respiration governed by the climatic regime? Soil Biol Biochem 22:525–532

Karaca A, Baran A, Kaktanir K (2000) The effect of compaction on urease enzyme activity, carbon dioxide evaluation, and nitrogen mineralization. Turk J Agric For 24:437–441

Ladd JN (1985) Soil enzymes. In: Vaughan D, Malcolm RE (eds) Soil organic matter and biological activity. Martinus Nijhoff, Boston, pp 175–221

Lee KE, Pankhurst CE (1992) Soil organisms and sustainable productivity. Aust J Soil Res 30:855–892

Megharaj M (2002) Heavy pesticide use lowers soil health. Kondinin Landcare Group Magazine, Farming Ahead 121:37–38

Megharaj M, Kantachote D, Singleton I, Naidu R (2000) Effects of long-term contamination of DDT on soil microflora with special reference to soil algae and algal metabolism of DDT. Environ Pollut 109:35–42

Michael AC, Turgeona AJ (1978) Microbial activity in soil and litter underlying bandane and calcium arsenate-treated turfgrass. Soil Biol Biochem 10:181–186

Nannipieri P (1994) The potential use of soil enzymes as indicators of productivity, sustainability and pollution. In: Pankhurst CE, Doube BM, VVSR G, Grace PR (eds) Soil biota: management in sustainable farming systems. CSIRO, Melbourne, pp 238–244

Nannipieri P, Grego S, Ceccanti B (1990) Ecological significance of biological activity in soil. In: Bollag JM, Stotzky G (eds) Soil biochemistry, vol 6. Marcel Dekker, New York, pp 293–355

Pankhurst C (2006) Effects of pesticides used in sugarcane cropping systems on soil organisms and biological functions associated with soil health. A report prepared for the sugar yield decline joint venture. Adelaide, pp 1–39

Pankhurst CE, Hawke BG, McDonald HJ, Kirkby CA, Buckerfield JC, Michelsen P, O'Brien KA, Gupta VVSR, Doube BM (1995) Evaluation of soil biological properties as potential bioindicators of soil health. Aust J Exp Agric 35:1015–1028

Paul EA, Mclaren AD (1975) Biochemistry of soil subsystem. In: Paul EA, Mclaren AD (eds) Soil biochemistry, vol 3. Marcel Dekker, New York, pp 1–36

Powlson DS, Brookes PC, Christensen BT (1987) Measurement of soil microbial biomass provides an early indication of changes in total soil organic matter due to straw incorporation. Soil Biol Biochem 19:159–164

Rai JPN (1992) Effect of long-term 2,4-D application on soil microbial populations. Biol Fertil Soils 13:427–431

Ramakrishnan B, Megharaj M, Venkateswarlu K, Naidu R, Sethunathan N (2010) The impacts of environmental pollutants on microalgae and cyanobacteria. Crit Rev Environ Sci Technol 40:699–821

Ramakrishnan B, Megharaj M, Venkateswarlu K, Sethunathan N, Naidu R (2011) Mixtures of environmental pollutants: effects of microorganisms and their activities in soils. Rev Environ Contam Toxicol 211:63–120

Ronnpagel K, Janben E, Ahlf W (1998) Asking for the indictor function of bioassays evaluating soil contaminations: are the bioassay results reasonable surrogates of effects on sol microflora? Chemosphere 36:1291–1304

Sannino F, Gianfreda L (2001) Pesticide influence on soil enzymatic activities. Chemosphere 45:417–425

Schäffer A (1993) Pesticide effects on enzyme activities in the soil ecosystem. In: Bollag JM, Stotzky G (eds) Soil biochemistry, vol 8. Marcel Dekker, New York, pp 273–340

Singh B, Walker A (2006) Microbial degradation of organophosphorus compounds. FEMS Microbiol Rev 30:428–471

Singh BK, Allan W, Denus JW (2002) Degradation of chlorpyrifos, fenamiphos and chlorothalonil alone and in combination and their effects on soil microbial activity. Environ Toxicol Chem 21:2600–2605

Soil Pollution (2017) Retrived from http://nsdl.niscair.res.in/jspui/bitstream-/123456789/990/1/Soil_Pollution.pdf. Accessed April 2017

Sparling GP (1985) The soil biomass. In: Vaughan D, Malcolm RE (eds) Soil organic matter and biological activity. Martinus Nijhoff/Dr. W. Junk, Dordrecht, pp 224–262

Tateno M (1988) Limitations of available substances for the expression of cellulase and protease activities in soil. Soil Biol Biochem 20:117–118

Taylor JP, Wilson B, Mills MS, Burns RG (2002) Comparison of microbial number and enzymatic activities in surface soils and subsoil using various techniques. Soil Biol Biochem 34:387–401

Trasar-Capeda C, Lieros MC, Seoane S, Gil-Sotres F (2000) Limitations of soil enzymes as indicators of soil pollution. Soil Biol Biochem 32:1867–1875

Tscherko D, Kandeler E (1999) Classification and monitoring of soil microbial biomass, N-mineralization and enzyme activities to indicate environmental changes. J Land Manage Food Environ 50:215–226

Tu CM (1995) Effect of five insecticides on microbial and enzymatic activities in sandy soil. J Environ Sci Health B30:289–306

Venkateswarlu K (1993) Pesticide interactions with cyanobacteria in soil and culture. In: Bollag JM, Stozky G (eds) Soil biochemistry, vol 8. Marcel Dekker, New York, pp 137–179

Voets JP, Meerschman P, Verstraete W (1974) Soil microbiological and biochemical effects of long-term atrazine applications. Soil Biol Biochem 6:149–152

Vrieze Jop de (2015) The littlest farmhands. Science 349:680–683

Waldrop MP, Balser TC, Firestone MK (2000) Linking microbial community composition to function in a tropical soil. Soil Biol Biochem 32:1837–1846

Wyszkowska J, Kucharski J (2004) Biochemical and physicochemical properties of soil contaminated with herbicide Triflurex 250 EC. Pol J Environ Stud 3:223–231

Xie X, Liao M, Huang C, Liu W, Abid S (2004) Effects of pesticides on soil biochemical characteristics of a paddy soil. J Environ Sci 16:252–255

Chapter 3
Selected Soils, Insecticides and Soil Enzymes

Soil Samples

Cotton is the predominantly grown crop in inherently very fertile agricultural fields of Nandyal. Under rain-fed conditions, millets, soybean, sorghum, pigeon pea, etc. are also grown in these fields, whereas a variety of other crops such as sugar cane, wheat, tobacco and citrus are cultivated under irrigated conditions. Soils, without or with a known history of insecticide (acephate or buprofezin) use, were collected to a depth of 12 cm from fields under cultivation of cotton. The collected soil samples were mixed, air-dried and sieved through a 2-mm mesh prior to use. The physico-chemical (Table 3.1) and microbiological characteristics were determined following the standard procedures. Soil pH was determined using an electrode in soil-water (1:1.25) slurry (Thomas 1996). Electrical conductivity of soil solution (1.0 g soil sample in 100 mL water) was determined using Elico conductivity meter. The method described by Johnson and Ulrich (1960) was employed for estimating 60% water-holding capacity. Total content of organic carbon and nitrogen in soil were quantified using Walkley-Black method (Nelson and Sommers 1996) and Micro-Kjeldahl's method (Jackson 1973), respectively. Populations of bacteria and fungi in both the soil samples (with and without insecticide use) were isolated and enumerated following serial dilution and plating method. The total bacterial population in soil without and with history of insecticide use was 3×10^8 and 2×10^9 CFU g^{-1} soil, respectively, while the fungal population in the corresponding soil samples was 6×10^4 and 2×10^5 CFU g^{-1} soil, respectively.

N.R. Maddela, K. Venkateswarlu, *Insecticides–Soil Microbiota Interactions*,
DOI 10.1007/978-3-319-66589-4_3

Table 3.1 Physico-chemical properties of the soil used

Characteristic	Value
pH (1:1.25 soil–water slurry)	8.2
Texture:	
Clay (%)	64
Silt (%)	15
Sand (%)	21
Electrical conductivity (μmhos cm^{-1})	0.24
60% water-holding capacity (mL 100 g^{-1})	45.6
Organic matter (%)	3.602
Total nitrogen (g kg^{-1} soil)	0.14
Available potassium (g K$_2$O kg^{-1} soil)	4.19
Available phosphorus (mg P$_2$O$_5$ g^{-1} soil)	4.25
Calcium	10–15%

Insecticides

Two insecticides, acephate (O,S-dimethyl acetyl phosphoramidothioate) and buprofezin (2-tert-butylimino-3-isopropyl-5-phenyl-1,3,5-thiadiazinan-4-one) were selected for the present investigation in view of their extensive and intensive use in Indian agriculture, in general, and Nandyal division, in particular, for control of major insect pests on cotton (Table 3.2). Stock solutions from commercial formulations of acephate (Hythene®, 75% SP) and buprofezin (Applaud®, 25% SC) were prepared in sterile distilled water for spiking soil samples at different concentrations.

Acephate is an organophosphate foliar spray insecticide of moderate persistence with residual systemic activity of about 10–15 days at the recommended rate of 292–584 g ha^{-1} (Table 3.3). Acephate is a linear compound with a molecular formula of $C_4H_{10}NO_3PS$ and a molecular weight of 183.17 (Fig. 3.1). It is very effective in controlling a wide range of biting and sucking insects, especially aphids in fruits, vegetables (e.g., potatoes and sugar beets), vine, cotton, rice, Bengalgram, hop cultivation and in horticulture (e.g., roses and chrysanthemum grown outdoors). It also controls leaf miners, lepidopterous larvae, sawflies and thrips on major plant species. Acephate is available in soluble powder, pressurized spray and granular formulations. Buprofezin belongs to thiadiazine insecticide group (Table 3.3), available in white suspension. It is a ring compound with an empirical formula of $C_{16}H_{23}N_3OS$ and a molecular weight of 305.4 (Fig. 3.2). It is extensively used on rice, citrus, mango, cotton, vegetables, ornamentals, grape and tea crops for controlling scale, mealy bug, jassids (leafhoppers), plant hoppers, silver leaf white flies at a dose of 250 g ha^{-1}.

Table 3.2 Use of acephate and buprofezin in Nandyal division, Kurnool district

| Month in 2009 | Insecticide | |
	Acephate (Hythene®, 75% SP) (kg)	Buprofezin (Applaud®, 25% SC) (liters)
January	1500	500
February	600	500
March	600	_a
April	400	–
May	100	–
June	200	–
July	500	–
August	800	300
September	1000	16,000
October	2200	11,000
November	2800	6000
December	1500	1000
Total per annum	12,200	35,300

[a]Not available

Table 3.3 Particulars of the insecticides used in the study

Feature	Acephate	Buprofezin
Chemical class	Organophosphate	Thiadiazine compound
Appearance	Colourless to white solid	White suspension
Molecular formula	$C_4H_{10}NO_3PS$	$C_{16}H_{23}N_3OS$
Commercial formulation	Hythene® (75% SP) from Hyderabad Chemicals Ltd., Hyderabad	Applaud® (25% SC) from Rallies India Ltd., Mumbai
Structure	Linear form	Ring form
Molecular weight	183.17	305.4
Usage dose	292–584 g ha^{-1}	250 g ha^{-1}
Crops applied	Bengal gram, cotton, rice, safflower, potatoes, sugar beets, hop cultivation, roses, chrysanthemums.	Rice, citrus, mango, cotton, vegetables, ornamentals, grape, tea.
Insect pests controlled	Aphids, bollworms, sawflies, thrips, stem borers, leaf folders, planthoppers, green leafhoppers.	Scale, mealy bug, jassids (leaf hoppers), planthoppers, silver leaf white flies.

Fertilizers

In easily available form to crops, nitrogen (N) is in ammonical form, and entire phosphate (P) and potash (K) are in water-soluble form. The mineral fertilizers such as urea, calcium perphosphate and potassium are used as a complex at a rate of 120 kg N h^{-1}, 80 kg P_2O_5 h^{-1} and 60 kg K_2O h^{-1}. The above complex was used

Fig. 3.1 Commercial formulation (*top*) and molecular formula (*bottom*) of acephate

$(C_4H_{10}NO_3PS)$

in the study as it is an ideal ratio of N, P and K particularly for cotton, rice, groundnut, chilies, soybean, potato and other commercial crops which require high phosphate initially.

Selected Soil Enzymes

Soil health is referred to as the balance among organisms within a soil, and between soil organisms and their environment (Brady and Weil 2008). Any disturbances in soil, as a consequence of detrimental effects of toxicants on soil biochemical activities, affects soil health which is of utmost importance for the survival of plants. Several characteristics of soil microorganisms such as viable counts, biomass, respiration, and enzyme activities have been largely considered as indices for biological assessment of soil ecotoxicity (Frankenberger and Dick 1983; Ramakrishnan et al. 2010). For instance, Insam et al. (1996) reported the use of soil enzymes as potential indicators of soil quality because of their susceptibility to marked changes due to soil contamination. Environmental pollutants of varying nature are known to exert differential effects on soil enzymes (Lethbridge et al. 1981; Ramakrishnan

Fig. 3.2 Commercial
formulation (*top*) and
molecular formula
(*bottom*) of buprofezin

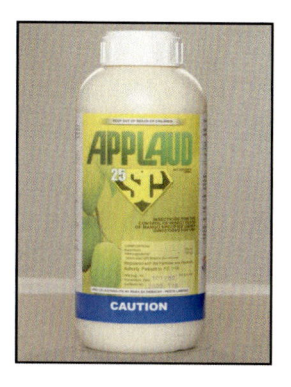

$(C_{16}H_{23}N_3OS)$

et al. 2010). Therefore, soil health has been assessed under the impact of the selected insecticides using the following soil enzymes, viz., cellulases, amylases, invertase, proteases, urease and phosphatases, as toxicity criteria.

Cellulases

Cellulose is the most abundant organic compound in the biosphere, comprising almost 50% of the biomass synthesized by photosynthetic fixation of CO_2 (Eriksson et al. 1990). Growth and survival of important microorganisms in most agricultural soils depends on carbon source contained in cellulose occurring in soils (Deng and Tabatabai 1994). However, for carbon to be released as an energy source for use by the microorganisms, cellulose in plant debris has to be degraded into glucose, cellobiose and high molecular weight oligosaccharides by cellulase enzymes (White 1982). It is apparent that cellulases are a group of enzymes that catalyze the degradation of cellulose, a polysaccharide built of β-1,4 linked glucose units (Deng and Tabatabai 1994). This group consists of endo-1,4-β-D-glucanase (EC 3.2.1.4), exo-1,4-D-glucanase (EC 3.2.1.155), and β-D-glucosidase (EC 3.2.1.21). It has been established that cellulases in soils are derived mainly from plant debris incorporated into soil, and that a limited amount of it may also originate from fungi and bacteria in soils (Richmond 1991). The activity of cellulases was indicated by the degradation of substrate like cellulose polymer of cellophane (Markus 1955), cellulose powder (Rawald et al. 1968), carboxymethyl cellulose (Kong and Domergues 1972), and its activity was measured by the method of Pancholy and Rice (1973) through appearance of reducing sugars measured spectrophotometrically. Nevertheless, the cellulase activity was potentially correlated with fungal and bacterial population in soil (Joshi et al. 1993). Furthermore, investigations indicated that activities of cellulases in agricultural soils are affected by several factors such as temperature, soil pH, water, oxygen, the chemical structure of organic matter and its location in soil horizon (Rubidge 1977), quality of organic matter/plant debris and soil mineral elements (Burns 1978), and trace elements from fungicides (Atlas et al. 1978).

In terms of soil fertility, processing and mineralization of plant and animal residues occurring in soil have great significance. Cellulases play a vital role in soil fertility by degrading cellulose materials. They decompose cellulose, either serially or synergistically, and also some related polysaccharides, and this biochemical activity is an important step of C-cycle mediated by soil microorganisms. Generally speaking, cellulase system of soil operates at the first step in the process that cellulose-C is mineralized to carbon dioxide and released into the atmosphere. It has been identified that the existence of cellulases, hemicellulases, and pectinases has potential application in agriculture for enhancing growth of crops and controlling plant diseases (Bhat 2000). For instance, cellulases and related enzymes from certain fungi are capable of degrading the cell wall of plant pathogens thereby controlling plant diseases (Bhat 2000). Many cellulolytic fungi in soil such as *Trichoderma* sp., *Geocladium* sp., *Chaetomium* sp., and *Penicillium* sp. were found to be good agents in improving plant growth and crop yields (Harman and Kubicek 1998). Furthermore, cellulases have also been identified in accelerating straw decomposition and increase in soil fertility (Han and He 2010). This is because of the fact that in a traditional agricultural practice, straw incorporation is an important strategy to improve soil quality and reduce dependence on commercial fertilizers (Tejada et al. 2008). Also, cellulolytic enzyme system in soil is essential in humification of plant residues which is an important process in the maintenance of soil health. By degrading insoluble cellulose materials, soil cellulases enhance the nutritional status even in mangrove communities (Behera et al. 2016). However, introduction of anthropogenic chemicals such as pesticides into soil may have long lasting effects on activities of soil cellulases and eventually soil health.

Amylases

Amylases are a group of extracellular enzymes which breakdown starch molecules (Ross 1976). During starch degradation by amylases, a molecule of water is also split into OH^- and H^+ ions which are bound to the exposed ends of the broken starch polymer; hence, the reaction is named *hydrolysis* (splitting of water molecule). It has long been known that amylases are widely distributed in plants and soils, and play a significant role in the breakdown of starch. There are two kinds of amylases namely, α-amylase (4-α-D-glucan glucanohydrolase, EC 3.2.1.1) and β-amylase (4-α-D-glucan maltohydrolase, EC 3.2.1.2) (Pazur 1965). Although several other enzymes are also involved in hydrolyzing starch, the major ones include α-amylase that converts starch like substrates to glucose and/or oligosaccharides, and β-amylase which hydrolyzes starch to maltose (Thoma et al. 1971). Generally speaking, several diverse products such as dextrins and much smaller polymers composed of glucose units are generated upon the hydrolysis of starch molecules by amylases. Such smaller molecules can then be easily transported into the microbial cells. Based on the mode of action of amylases, they are categorized into exo-acting, endo-acting and debranching enzymes. For instance, β-amylase is an exo-acting enzyme whereas α-amylase is endo-acting. Studies have shown that α-amylases are synthesized by plants, animals and microorganisms, while β-amylase is mainly synthesized by plants (Thoma et al. 1971). A number of these starch hydrolyzing enzymes are also important industrial amylases.

Although several microorganisms (fungi, yeasts, bacteria and actinomycetes) are implicated in producing amylases, these enzymes from fungal and bacterial sources have predominant applications in industrial sectors. There are also some unusual amylases found in acidophilic, alkalophilic, and thermoacidophilic bacteria (Boyer and Ingle 1972). High levels of amylase-producing fungal species include *Aspergillus niger*, *A. oryzae*, *Thermomyces lamuginosus*, and *Penicillium expansum*. In particular, *Bacillus* spp., produce a large variety of extracellular enzymes, of which amylases are of particular significance to the industry. Studies have, however, indicated that the roles and activities of α-amylase and β-amylase may be influenced by different factors ranging from cultural practices, type of vegetation, environment, and soil types (Ross and Roberts 1968).

Together with cellulases and invertase, amylases greatly influence the rate and the course of decomposition of plant material in soil (Pancholy and Rice 1972). For this reason, amylases are also considered as one of the important C-cycling enzymes. These enzymes have been reported from different soil organisms and different habitats. For instance, earthworms which are crucial in maintaining soil fertility through hydrolyzing carbohydrates also have the ability to digest various organics in soil such as leaf litters, roots, yeast, brown algae, and fungi (Prat et al. 2002). This activity is mainly due to the presence of α-amylases in the body wall of earthworms that digests raw starch (Mitsuhiro et al. 2008). Amylase activity has also been detected in some permafrost sediment samples wherein culturable microorganisms are absent (Vorobyova et al. 1997). Amylases from fungi are particularly active in mild acidic pH range and they readily degrade a wide variety of starch-containing substrates in acidic soils.

Invertase

Invertase (β-D-fructofuranoside fructohydrolase, EC 3.2.1.26), predominantly available in microorganisms, animals and plants (Kiss and Peterfi 1959), catalyzes the hydrolysis of sucrose to glucose and fructose. This enzyme brings about complete hydrolysis of sucrose either under acid or alkaline conditions (Splading 1979).

Soil invertase received special attention because its substrate, sucrose, is one of the most abundant soluble sugars in plants. Invertase activities have also been observed in forest ecosystems (Frankenberger and Johanson 1983) as it is partially responsible for the breakdown of forest floor material. It can also degrade particularly water-soluble plant materials in soils (Ross 1983). Activity of invertase in soils is partly associated with light organic fractions. In general, invertase is bound to a variety of soil fractions such as organic fraction of high density, clay minerals and microbial biomass (Stemmer et al. 1998).

In soil ecosystem, invertase is affected by several physical and chemical factors. Soil pH has little effect on soil invertase activity, especially when soil pH is 4–9 (Frankenberger and Johanson 1982). When soil pH is very low or high, the C:N ratio has also been shown to affect the soil invertase (Ross 1973). A study was conducted in China focusing on urbanization and its impact on soil health wherein different statistical tools were used to analyze the data (Shi et al. 2008). According to the correlation analysis, soil invertase activity was significantly affected by soil organic matter but not by other soil physico-chemical variables. Path analysis showed that soil organic matter had a positive direct effect on soil invertase activity, while total N had a negative direct effect. On the other hand, total P, alkali-hydrolysable N and clay fractions had a positive indirect effect on invertase via soil organic matter. Soil available P, pH, and EC exerted no direct or indirect effects on invertase activity. Clearly, these observations indicate that the activity of soil invertase is highly influenced by both physical and chemical parameters of soil as well as soil management practices.

Proteases

Proteases (peptidyl peptide hydrolases, EC 3.4) are widely distributed among soils and show a wide range of activities (Ladd and Butler 1972). In fact, proteases reach the soil from different sources including plants, animal excrements and microorganisms. These extracellular enzymes are involved in the initial hydrolysis of protein components of organic nitrogen to simple amino acids. Proteases in soils hydrolyze not only added protein, but also native soil proteins (Kiss et al. 1975). In general, soil microorganisms produce proteases to recycle soil organic matter that ensures microbial nutrition. Thus, hydrolytic degradation of proteins is an important process in microbial N-cycling in many ecosystems. Proteolysis is thus considered to be a rate-limiting step during N mineralization in soils (Weintraub and Schimel 2005) because mineralization of primary phase of protein is much slower than that of amino acid mineralization (Jan et al. 2009). Therefore, N mineralziation mediated by proteases is an important process regulating the amount of plant available N and plant growth.

In addition to N-cycling, proteases perform several other functions in soil. There is a significant role of proteases in the interactions of soil organisms via cleavage of the cell wall proteins which greatly helps in the microbial inter-cellular interactions in soil. All anti-fungal proteases of bacterial origin and alkaline serine proteases of nematophagous or entomopathogenic bacteria and fungi come under this category (Chang et al. 2007). On the other hand, keratinolytic serine proteases, produced by different soil bacteria and fungi play an important role in recycling keratinous residues (Gradišar et al. 2005). Furthermore, proteases help in the survival of microorganisms under unfavorable conditions. For example, periplasmic heat shock proteins of *Escherichia coli* possess proteolytic activities that enable its survival even at high temperatures (Kim et al. 1999). Certain proteases namely serine endopeptidases are also involved in the development, pathogenesis and biocontrol of microorganisms (Pöll et al. 2009).

Except certain thermo-tolerant ligninolytic fungi and some mycobacteria and clostridia, most soil microorganisms express proteolytic activities (Cruz et al. 2012). Key microbial proteases include serine alkaline peptidases, subtilisin-like peptidases and subtilisins which are secreted by certain bacteria and fungi. The pH optimum for most microbial proteases ranges from 3.0 to 12.0 (Singh et al. 2011), while most microorganisms isolated from soils possess thermophilic (40–60 °C) and alkaliphilic (pH 8–9) proteases (Shankar et al. 2011). It has also been identified that the proteases reaching soil from different sources have different kinetic properties (Valerie et al. 2013). For instance, molecular weights of plant and soil proteases are 175 and 75 kDa, respectively.

Generally, proteases exist in colloidal form by associating with various biotic and abiotic components in soil, including dividing and dormant cells, cell debris, clay minerals, humic colloids and the soil aqueous phase (Burns 1982). Various criteria are used to classify proteases such as type of reaction catalyzed, the active site functional group, molecular structure and evolutionary relationships (Landi et al. 2011). Nontheless, the activity of proteases in soil is known to be affected by several biotic and abiotic factors since they are extracellular in nature.

Urease

Urease (urea amidohydrolase, EC 3.5.1.5) was first reported by Rotini (1935), and is considered vital in the regulation of N supply to plants after urea fertilization since urease is responsible for the hydrolysis of urea fertilizer applied to soil to yield NH_3 and CO_2 with the concomitant rise in soil pH (Andrews et al. 1989). This, in turn, results in a rapid N loss to the atmosphere through NH_3 volatilization (Fillery et al. 1984). Urease activity in soils has therefore received considerable attention. Soil urease originates mainly from plants (Polacco 1977) and microorganisms, and released as both intra- and extra-cellular enzymes (Mulvaney and Bremner 1981). Significant fraction of extracellular ureases is well stabilized by organic and mineral soil colloids. This type of immobilization greatly protects urease from proteases, since urease is highly susceptible to proteolytic enzymes (Zantua and Bremner 1977). Thus, urease is one of the most important enzymes in the nitrogen cycling, and is acknowledged as very sensitive to pollution (Megharaj et al. 1998).

There are many factors that influence the activity of urease in soil such as cropping history, organic matter content, soil depth, amendments, heavy metals, temperature, etc. (Tabatabai 1977). Effects of urban soils on urease activity were studied and analyzed using such statistical methods as correlation and path analysis (Shi et al. 2008). For instance, soil urease activity was closely and positively correlated to several soil physico-chemical properties such as soil total N, total P, alkali-hydrolysable N and physical clay. According to path analysis, soil organic matter and alkali-hydrolysable N had positive direct effects on urease activity, while total N had a positive indirect effect via soil organic matter. Thus, with an increase in soil organic matter, total N inhibited urease activity in urban soils. Moreover, soil available P, pH and EC had no impact on urease activity.

Acid and Alkaline Phosphatases

The term phosphatases in soil is used to describe a group of enzymes that are responsible for the hydrolytic cleavage of a variety of ester-phosphate bonds of organic phosphates and anhydrides of orthophosphoric acid (H_3PO_4) into organic phosphate (Schmidt and Laskowski 1961). With their predominant occurrence in bacteria to mammals, phosphatases indicate their importance in fundamental biochemical processes (Posen 1967). In soil ecosystems, these enzymes are believed to play a critical role in P cycles (Speir and Ross 1975) since evidence shows that they are correlated to P stress and plant growth. Both acid phosphatases (phosphate-monoester phosphohydrolase (alkaline optimum), EC 3.13.1) and alkaline phosphatases (phosphate-monoester phosphohydrolase (acid optimum), EC 3.1.3.2) particularly hydrolyze the ester bonds binding to P and C (C-O-P ester bonds) in organic matter. During the process, inorganic P is released from organically-bound P such as leaf litter, dead root systems, and other organic debris without a crucial role in phosphorous acquisition of plants and microorganisms (Schneider et al. 2001). There is a good response of acid phosphatases to signal that indicates phosphorous deficiency in soil. Such signaling system increases the secretion of acid phosphatases from plant roots. This enhances the solubilization and remobilization of phosphate which makes the availability of phoshophorous to plants under P-stressed conditions (Karthikeyan et al. 2002). Likewise, apart from being good indicators of soil fertility, phosphatase enzymes play key roles in soil system (Dick and Tabatabai 1992).

Acid and alkaline phosphatases are exoenzymes and may be protected from degradation by adsorption to clays or to humic substances (Skujins 1976). According to path analysis, soil organic matter, total N, alkali-hydrolysable N and clay did not show any significant direct or indirect effect on phosphatase activity in urban soils (Shi et al. 2008). However, simple correlation analysis showed that phosphatase activity was closely and positively correlated with total N, total P, alkali-hydrolysable N and clay. Even though soil pH had a significant negative direct effect on phosphatase activity, it was counteracted by other soil properties

(Shi et al. 2008). Thus, as described earlier for other enzymes, phosphatases are also highly sensitive to soil treatments during such agricultural management practices as pesticide applications.

Experimental Setup

Aliquots (0.5 mL) from stock solutions of acephate and buprofezin, prepared in distilled water, were applied with 0.1 mL pipette to the surface of 5 g soil samples contained in test tubes (25 × 200 mm) as followed earlier by Lethbridge and Burns (1976). The final concentrations (w/w basis) of each insecticide included were 2.5, 5.0, 7.5 and 10.0 $\mu g\ g^{-1}$ soil, and these levels correspond to 0.25, 0.5, 0.75 and 1.0 kg ha^{-1}, respectively (Anderson 1978). These concentrations were chosen because of the fact that field application doses of the selected insecticides range from 0.3 to 0.6 kg ha^{-1}. Soil samples in a set were treated with different concentrations of the insecticides. Besides insecticide treatment, soil samples in another set were amended with N-P-K fertilizer at a rate of 120-80-60 kg ha^{-1}, respectively. The soil samples receiving only 0.5 mL distilled water served as controls. Soil samples in each set received an insecticide once, twice (second application after 15 days of first application), or thrice (third application after 30 days of first application). All the treatments including controls were maintained at 60% water-holding capacity, and incubated in the laboratory at 28 ± 4 °C. After 3 days of incubation, triplicate soil samples included for single application of the insecticides were withdrawn for the assay of cellulases (Pancholy and Rice 1973), amylases (Cole 1977; Tu 1981a, b), invertase (Tu 1982), proteases (Speir and Ross 1975), urease (Zantua and Bremner 1975) and acid phosphatase (Tabatabai and Bremner 1969), and alkaline phosphatase (Tabatabai and Bremner 1969; Eivazi and Tabatabai 1977) as described in the following chapters. Likewise, enzyme activities in soil samples were also determined separately 3 days after two (15 days of incubation) and three (30 days of incubation) repeated applications of the insecticides.

In another study, aqueous solutions of commercial formulations of the two insecticides, acephate and buprofezin, were added to 5 g portions of the soil in test tubes (25 × 200 mm) to get insecticide combinations, acephate + buprofezin with graded concentrations (2.5–10 $\mu g\ g^{-1}$ soil) as described earlier by Gundi et al. (2005) and Khajepour et al. (2012). Also, in one set, soil samples that received insecticide combinations were amended with fertilizers at a rate of 120-80-60 kg ha^{-1} of N-P-K, respectively. Soil samples without the addition of insecticides served as controls. All the tubes were incubated in the dark at room temperature (28 ± 4 °C) with 60% water-holding capacity. After 3 days of incubation, triplicate soil samples were withdrawn to determine the activities of cellulases, amylases, invertase, proteases, urease, acid and alkaline phosphatases in soil samples.

The per cent inhibition values for the enzymes were calculated relative to the activity in untreated controls. The data on interaction effects for the insecticide combinations employed were analyzed by the multiplicative survival model as outlined by Stratton (1983). The expected interaction responses for the insecticide combinations were calculated using the formula:

$$E = X + \left[(100 - X) / 100 \right] \times Y$$

where, E = the expected additive effect of the mixture, X = % inhibition due to component A alone, and Y = % inhibition due to component B alone. The mean ratios between actual inhibition and expected inhibition significantly greater and less than 1.0 indicated synergism and antagonism, respectively, while an additive effect occurred when the actual and expected inhibitions did not differ significantly (Stratton 1984).

Statistical Analysis

The statistical analyses were performed with IBM-SPSS program (SPSS, Inc., Chicago, USA). Significant differences among insecticide doses, number of applications, and N-P-K amendments were compared using Duncan's new multiple range test.

References

Anderson JR (1978) Pesticide effects on non-target soil microorganisms. In: Hill IR, Write SJL (eds) Pesticide microbiology. Academic, London, pp 313–533

Andrews RK, Blakeley RL, Zerner B (1989) Urease: A. Ni (II) metalloenzyme. In: Lancester JR (ed) The bioinorganic chemistry of nickel. V.C.H. Publishers, New York, pp 141–166

Atlas RM, Pramer D, Bartha R (1978) Assessment of pesticide effects on non-target soil microorganisms. Soil Biol Biochem 10:231–239

Behera BC, Sethi BK, Mishra RR, Dutta SK, Thatoi HN (2016) Microbial cellulases – diversity & biotechnology with reference to mangrove environment: a review. J Genet Eng Biotechnol. http://dx.doi.org/10.1016/j.jgeb.2016.12.001

Bhat MK (2000) Cellulases and related enzymes in biotechnology. Biotechnol Adv 18:355–383

Boyer EW, Ingle MB (1972) Extracellular alkaline amylase from *Bacillus* sp. J Bacteriol 110:992–1000

Brady NC, Weil RR (2008) Soil water: characteristics and behavior. In: Brady NC, Weil RR (eds) The nature and properties of soils. Prentice Hall, New Jersey, pp 177–217

Burns RG (1978) Soil enzymes. Academic, New York, p 370

Burns RG (1982) Enzyme activity in soil: location and a possible role in microbial ecology. Soil Biol Biochem 14:423–427

Chang WT, Chen YC, Jao CL (2007) Antifungal activity and enhancement of plant growth by *Bacillus cereus* grown on shellfish chitin wastes. Bioresour Technol 98:1224–1230

Cole MA (1977) Lead inhibition of enzyme synthesis in soil. Appl Enviorn Microbiol 33:262–268

Cruz RMG, Rivera-Rios JM, Téllez-Jurado A, Gálvez APM, Mercado-Flores Y, Arana-Cuenca AA (2012) Screening for thermotolerant ligninolytic fungi with laccase, lipase, and protease activity isolated in Mexico. J Environ Manag 95:S256–S259

Deng SP, Tabatabai MA (1994) Cellulase activity of soils. Soil Biol Biochem 7:5–8

Dick WA, Tabatabai MA (1992) Potential uses of soil enzymes. In: Metting FB Jr (ed) Soil microbial ecology, applications in agricultural and environmental management. Marcel Dekker, New York, pp 95–127

Eivazi F, Tabatabai MA (1977) Phosphatase in soil. Soil Biol Biochem 9:167–172

Eriksson KEL, Blancbette RA, Ander P (1990) Biodegradation of cellulose. In: Eriksson KEL, Blanchette RA, Ander P (eds) Microbial and enzymatic degradation of wood and wood components. Springer, New York, pp 89–180

Fillery IRP, Simpson JR, De Datta SK (1984) Influence of field environment and fertilizer management on ammonia loss from flooded rice. Soil Sci Soc Am J 48:914–920

Frankenberger WT Jr, Dick WA (1983) Relationships between enzyme activities and microbial growth and activity indices in soil. Soil Sci Soc Am J 47:945–951

Frankenberger WT, Johanson JB (1982) Effect of pH on enzyme stability in soils. Rom Biotech Lett 14:433–437

Frankenberger WT, Johanson JB (1983) Factors affecting invertase activity in soils. Plant Soil 74:313–323

Gradišar H, Friedrich J, Križaj I, Jerala R (2005) Similarities and specificities of fungal keratinolytic proteases: comparison of keratinases of *Paecilomyces marquandii* and *Doratomyces microsporus* to some known proteases. Appl Environ Microbiol 71:3420–3434

Gundi VAKB, Narasimha G, Reddy BR (2005) Interaction effects of insecticides on microbial populations and dehydrogenase activity in black clay soil. J Sci Environ Health B40:269–283

Han W, He M (2010) The application of exogenous cellulase to improve soil fertility and plant growth due to acceleration of straw decomposition. Bioresour Technol 101:3724–3731

Harman GE, Kubicek CP (1998) *Trichoderma* and *Gliocladium*: enzymes, biological control and commercial applications, vol 2. Taylor & Francis, London

Insam H, Ranger A, Henrich M, Hitzl W (1996) The effect of grazing on soil microbial biomass and community on alpine pastures. Phyton-Horn 36:205–216

Jackson ML (1973) The text book of soil chemical analysis. Prentice-Hall, Engle Wood Cliffs, p 187

Jan TM, Roberts P, Tonheim SK, Jones DL (2009) Protein breakdown represents a major bottleneck in nitrogen cycling in grassland soils. Soil Biol Biochem 41:2272–2282

Johnson CM, Ulrich A (1960) Determination of moisture in plant tissues. California agricultural bulletin no. 766. In: Wilde SA (ed) Soil and plant analysis for tree culture. Obrtage Publishing Co, Oxford/Bombay, pp 112–115

Joshi SR, Sharma GD, Mishra RR (1993) Microbial enzyme activities related to litter decomposition near a highway in a subtropical forest of north East India. Soil Biol Biochem 22:51–55

Karthikeyan AS, Varadarajan DK, Mukatira UT, D'Urzo MP, Damaz B, Raghothama KG (2002) Regulated expression of *Arabidopsis* phosphate transporters. Plant Physiol 130:221–233

Khajepour S, Izadi H, Asari MJ (2012) Evaluation of two formulated chitin synthesis inhibitors, hexaflumuron and lufenuron against the raisin moth, *Ephestia figulilella*. J Insect Sci 12:102

Kim KI, Park SC, Kang SH, Cheong GW, Chung CH (1999) Selective degradation of unfolded proteins by the self-compartmentalizing HtrA protease, a periplasmic heat shock protein in *Escherichia coli*. J Mol Biol 294:1363–1374

Kiss S, Peterfi Jr (1959) Biologia 2:179. (Cinted in: Tu CM (1982)) Influence of pesticides on activities of invertase, amylase and level of adenosine tri phosphate in organic soil. Chemosphere 11:909–914

Kiss S, Sstetanio G, Diagnan-Vulanddra M (1975) Soil enzymology in Romania Part II. Contribute Botanica Cluj 197–207

Kong KT, Domergues Y (1972) Limiting cellulolysis in organic soils. II. Study of soil. Eur J Soil Biol (Du Revue d'Ecologie et de Biologie sol) 9:629–640

Ladd JN, Butler JHA (1972) Short-term assays of soil proteolytic enzyme activities using proteins and dipeptide derivatives as substrates. Soil Biol Biochem 4:19–30

Landi L, Renella G, Giagnoni L, Nannipieri P (2011) Activities of proteolytic enzymes R.P. Dick (Ed.), Methods of soil enzymology, Soil Science Society of America, Madison 247–273

Lethbridge G, Burns RG (1976) Inhibition of soil urease by organophosphorus insecticides. Soil Biol Biochem 8:99–102

Lethbridge G, Bull AT, Burns RG (1981) Effects of pesticides on 1,3-β-glucanase and urease activities in soils in the presence and absence of fertilizers, lime and organic materials. Pestic Sci 12:147–155

Markus L (1955) Determination of carbohydrates from plant materials with anthrone reagent: assay of cellulase activity in soil and farmyard manure. Agrochem Soil Sci 4:207–216

Megharaj M, Singleton I, McClure N (1998) Effect of pentachlorophenol pollution towards micro-algae and microbial activities in soil from a former timber processing facility. Bull Environ Contam Toxicol 61:108–115

Mitsuhiro U, Tomohiko A, Masami N, Kazutaka M, Kuniyo I (2008) Purification and characterization of novel raw-starch-digesting and cold-adapted α-amylases from *Eisenia foetida*. Comp Biochem Physiol B150:125–130

Mulvaney RL, Bremner JM (1981) Control of urea transformation in soils. In: Paul EA, Ladd JN (eds) Soil biochemistry, vol 5. Marcel Dekker, New York, pp 153–196

Nelson DW, Sommers LE (1996) Total carbon, organic and organic matter. In: Sparks DL (ed) Methods of soil analysis, part 3, Soil Science Society of America book series, vol 5. American Society of Agronomy and Soil Science Society of America, Madison, pp 961–1010

Pancholy SK, Rice EL (1972) Effect of storage conditions on activities of urease, invertase, amylase and dehydrogenase in soil. Soil Sci Am Proc 36:536–537

Pancholy SK, Rice EL (1973) Soil enzymes in relation to old field succession: amylase, cellulase, invertase, dehydrogenase and urease. Soil Sci Soc Am Proc 37:47–50

Pazur JH (1965) Enzymes in the synthesis and hydrolysis of starch. In: Whistler R, Paschall EF (eds) Starch: chemistry and technology, fundamental aspects, vol 1. Academic, New York, pp 133–175

Polacco JC (1977) Is nickel a universal component of plant ureases? Plant Sci Lett 10:249–255

Pöll V, Denk U, Shen HD, Panzani RC, Dissertori O, Lackner P, Hemmer W, Mari A, Crameri R, Lottspeich F, Rid R, Richter K, Breitenbach M, Simon-Nobbe B (2009) The vacuolar serine protease, a cross-reactive allergen from *Cladosporium herbarum*. Mol Immunol 46:1360–1373

Posen S (1967) Alkaline phosphatase. Ann Intern Med 67:183–203

Prat P, Charrier M, Deleporte S, Frenot Y (2002) Digestive carbohydrases in two epigenic earthworm species of the Kerguelen Islands (Subantarctic). Pedobiologia 46:417–427

Ramakrishnan B, Megharaj M, Venkateswarlu K, Naidu R, Sethunathan N (2010) The impacts of environmental pollutants on microalgae and cyanobacteria. Crit Rev Environ Sci Technol 40:699–821

Rawald LW, Domke K, Stohr G (1968) Studies on relations between humus quality and microflora of soil. Pedobiologia 7:375–380

Richmond PA (1991) Occurrence and functions of native cellulose. In: Haigler CH, Weimer PJ (eds) Biosynthesis and biodegradation of cellulose. Marcel Dekker, New York, pp 5–23

Ross DJ (1973) Some enzyme and respiratory activities of tropical soils from new hebrides. Soil Biol Biochem 5:559–567

Ross DJ (1976) Invertase and amylase activities as ryegrass and white clover plants and their relationships with activities in soils under pasture. Soil Biol Biochem 8:351–356

Ross DJ (1983) Invertase and amylase activities as influenced by clay minerals, soil-clay fractions and topsoils under grassland. Soil Biol Biochem 15:287–293

Ross DJ, Roberts HS (1968) A study of activities of enzymes hydrolysis sucrose and starch and of oxygen uptake in a sequence of soils under tussock grassland. J Soil Sci 19:186–196

Rotini OT (1935) La Transformazione enzimatica dell'urea nel terreno. Ann Labor Ric Ferm Spallanrani 3:143–154

Rubidge T (1977) The effect of moisture content and incubation temperature upon the potential cellulase activity of John Innes no.1 soil. Int Biodeterior Biodegrad 13:39–44

Schmidt G, Laskowski M Sr (1961) Phosphate ester cleavage (survey). In: Boyer PD, Lardy H, Myrback K (eds) The enzymes, 2nd edn. Academic, New York, pp 3–35

Schneider K, Turrion MB, Grierson BF, Gallardo JF (2001) Phosphatase activity, microbial phosphorous, and fine root growth in forest soil in the sierra de Gata, western central Spain. Boil Fertil Soils 34:151–155

Shankar S, Rao M, Laxman S (2011) Purification and characterization of an alkaline protease by a new strain of *Beauveria* sp. Process Biochem 46:579–585

Shi ZJ, Lu Y, Xu ZG, Fu SL (2008) Enzyme activities of urban soils under different land use in the Shenzhen city, China. Plant Soil Environ 54:341–346

Singh SK, Singh SK, Tripathi VR, Khare SK, Garg SK (2011) A novel psychrotrophic, solvent tolerant *Pseudomonas putida* SKG-1 and solvent stability of its psychro-thermoalkalistable protease. Process Biochem 46:1430–1435

Skujins J (1976) Extracellular enzymes in soil. CRC Crit Rev Microbiol 4:383–421

Speir TW, Ross DJ (1975) Effects of storage on the activities of protease, urease, phosphatase and sulphatase in three soils under pasture. NZ J Sci 18:231–237

Splading BP (1979) Effects of divalent metal chlorides on respiration and extractable enzymatic activities of Douglas-Fir needle litter. J Environ Qual 8:105–109

Stemmer M, Gerzabek MH, Kandeler E (1998) Organic matter and enzyme activity in particle-size fractions of soils obtained after low-energy sonication. Soil Biol Biochem 30:9–17

Stratton GW (1983) Interaction effects of permethrin and atrazine combinations towards several non-target microorganisms. Bull Environ Contam Toxicol 31:297–303

Stratton GW (1984) Effects of the herbicide atrazine and its degradation products, alone and in combination on phototrophic microorganisms. Arch Environ Contam Toxicol 13:35–42

Tabatabai MA (1977) Effects of trace elements on urease activity in soils. Soil Biol Biochem 9:9–13

Tabatabai MA, Bremner JM (1969) Use of *p*-nitrophenyl phosphate for assay of soil phosphatase activity. Soil Biol Biochem 1:301–307

Tejada M, Gonzalez JL, García-Martínez AM, Parrado J (2008) Application of a green manure and green manure composted with beet vinasse on soil restoration: effects on soil properties. Bioresour Technol 99:4949–4957

Thoma JK, Spradlin JE, Dygert S (1971) Plant and animal amylases. In: Boyer PD (ed) The enzymes, vol 5, 3rd edn. Academic, New York, pp 115–189

Thomas GW (1996) Soil pH and soil acidity. In: Sparks DL (ed) Methods of soil analysis, part 3, Soil Science Society of America book series, vol 5. American Society of Agronomy and Soil Science Society of America, Madison, pp 475–490

Tu CM (1981a) Effect of pesticides on activity of enzymes and microorganisms in a clay loam soil. J Environ Sci Health B16:179–191

Tu CM (1981b) Effect of some pesticides on enzyme activities in an organic soil. Bull Environ Contam Toxicol 27:109–114

Tu CM (1982) Influence of pesticides on activities of amylase, invertase and level of adenosine triphosphate in organic soil. Chemosphere 2:909–914

Valerie V, Klement R, Pavel F (2013) Proteolytic activity in soil: a review. Appl Soil Ecol 70:23–32

Vorobyova E, Soina V, Gorlenko M, Minkovskaya N, Zalinova N, Mamukelashvili A, Gilichinsky D, Rivkina E, Vishnivetskaya T (1997) The deep cold biosphere: facts and hypothesis. FEMS Microbiol Rev 20:277–290

Weintraub MN, Schimel JP (2005) Seasonal protein dynamics in Alaskan arctic tundra soils. Soil Biol Biochem 37:1469–1475

White AR (1982) Visualization of cellulases and cellulose degradation. In: Brown RM Jr (ed) Cellulose and other natural polymer systems: biogenesis, structure, and degradation. Plenum, New York, pp 489–509

Zantua MI, Bremner JM (1975) Comparison of methods of assaying urease activity in soils. Soil Biol Biochem 7:291–295

Zantua MI, Bremner JM (1977) Stability of urease in soils. Soil Biol Biochem 9:135–140

Chapter 4
Impact of Acephate and Buprofezin on Soil Cellulases

Assay of Cellulases in Soil

The activity of cellulases in soil samples was assayed by the method described by Pancholy and Rice (1973). Triplicate samples of soil (5 g) were withdrawn after desired intervals, placed in 50 mL Erlenmeyer flasks, and 0.5 mL of toluene was added. Contents in the flasks were mixed thoroughly, and 10 mL of 0.5 M acetate buffer (pH 5.9) was added after 15 min followed by the addition of 10 mL of 1% carboxy methyl cellulose (CMC). After 30 min of incubation, 50 mL of distilled water was added to this mixture. Then the suspension was filtered by Whatman No.1 filter paper, and volume of the filtrate was made up to 100 mL with distilled water. A suitable aliquot of the supernatant was treated with 2 mL of alkaline copper reagent (Nelson 1944), tubes were placed in boiling water bath for 10 min and cooled to room temperature. The mixture was treated with 1.0 mL of arsenomolybdate reagent followed by the addition of 5 mL of distilled water. The blue-colored complex developed was read at 620 nm in a spectrophotometer. Glucose was used as a standard. The activity of cellulases was expressed as milligrams of glucose released per g of soil per 30 min (mg glucose g^{-1} soil 30 min^{-1}).

Nontarget Effects on Cellulases

Involving a single or repeated applications ranging from 2.5 to 10 $\mu g\ g^{-1}$ soil, the nontarget effects of acephate and buprofezin toward soil cellulases were assessed by quantifying glucose released from the substrate, CMC. Acephate application to the soil particularly at 5 $\mu g\ g^{-1}$ soil level greatly increased (240–350%) the enzyme activity (Fig. 4.1a). Again, two repeated applications were more effective in enhancing the enzyme activity. On the other hand, single application of buprofezin, at 2.5–7.5 $\mu g\ g^{-1}$

© Springer International Publishing AG 2018
N.R. Maddela, K. Venkateswarlu, *Insecticides–Soil Microbiota Interactions*,
DOI 10.1007/978-3-319-66589-4_4

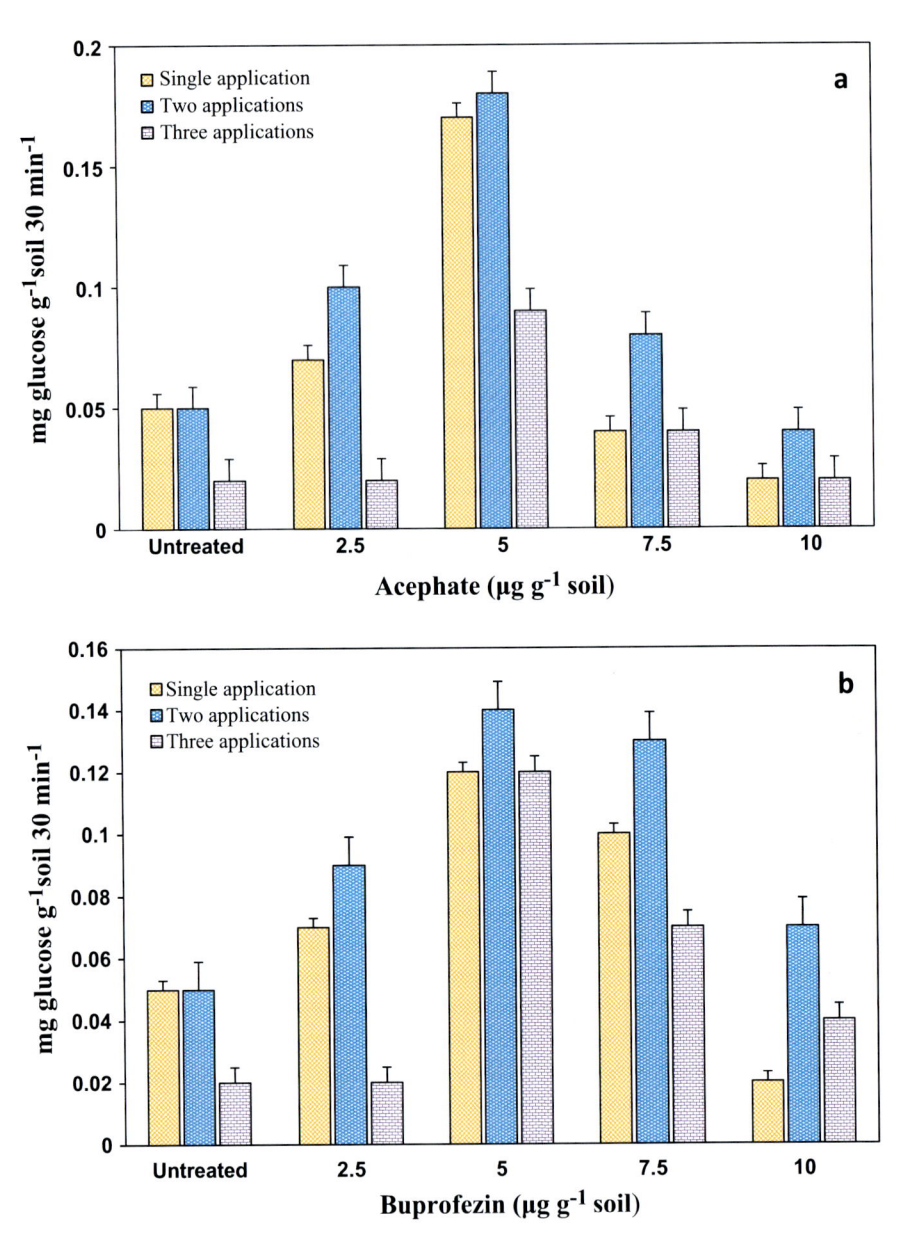

Fig. 4.1 Effect of single and repeated applications of (**a**) acephate and (**b**) buprofezin on activity of cellulases in soil. Error bars represent standard deviations (n = 3)

soil, resulted in a significant increase (40–140%) in enzyme activity (Fig. 4.1b). Even two or three applications at these concentrations showed pronounced activity of cellulases. Interestingly, there was a significant increase in enzyme activity after two repeated applications of the insecticide. However, the highest concentration of $10~\mu g~g^{-1}$ soil was significantly toxic to the enzyme. Overall, the impact of buprofezin toward the activity of cellulases was similar to that of acephate.

Although similar results have been reported by many researchers with several OP insecticides, virtually there is no study revealing the impact of thiadiazine related compounds on cellulases or even populations of microflora in soil. Gundi et al. (2007) reported that the soil treated with monocrotophos, up to $25~\mu g~g^{-1}$, was either stimulatory or nontoxic, but soil cellulase activities were adversely affected at higher rates of this insecticide. Similarly, cellulase activity was reported to increase up to 50% when the nematicide fenamiphos at $18.6~kg~ha^{-1}$ was applied to a fine silty montmorillonite soil in the field under grass-clover pasture (Ross et al. 1984). However, at a concentration of $930~mg~kg^{-1}$, fenamiphos had a deleterious effect on cellulase activity, and was reduced by 48% after 62 days of treatment under laboratory conditions (Ross and Speir 1985). Activity of cellulase in soil was adversely affected by 1% dimethoate in the initial period of incubation (Begum and Rajesh 2015). In three soils treated with dimethoate, activities of cellulases were decreased by 6–50% from day 0 to 45. However, from day 60 onwards dimethoate-treated soils have shown higher activity than that of control soils (Begum and Rajesh 2015). One possible reason for higher cellulase activities in treated soils after days 60 over controls was due to restoration of microflora utilizing dimethoate. Shittu et al. (2004) suggested that there was an initial rise in bacterial population in soil samples treated with diazinon and thereafter population was declined, while fungal population decreased considerably.

Selecron, an organophosphorus insecticide, at 10 and 50 ppm significantly decreased respiration, mycelial protein, extracellular protein and mycelial dry weight of *Aspergillus fumigatus*, *A. terreus* and *Myceliphthora thermophila* when grown at 45 °C (Omar et al. 1993). C_x and C_1 cellulases of tested fungi were significantly affected by this insecticide. However, C_1 cellulase of *A. fumigatus* was slightly increased. Lodhi et al. (2000) reported that cellulase activity was not much affected, although an increase of 18.5% was observed at $1.6~\mu g~g^{-1}$ soil of baythroid. At the highest concentration of baythroid, however, cellulase activity was reduced by 25.9%. In contrast, insecticides like quinalphos and cypermethrin (Gundi et al. 2007), fungicides such as tridemorph, captan (Srinivasulu and Rangaswamy 2006), pyrazofos and propiconazole (Omar and Abd-Alla 2000) significantly enhanced soil cellulase activity at lower doses, whereas higher rates of these pesticides were either innocuous or toxic to the enzyme activities. The insecticide, cartap hydrochloride, did not affect soil cellulase activity (Endo et al. 1982).

One of the rate limiting enzymes in microbial degradation of cellulose to glucose is β-glucosidase. Its response against chlorpyrifos and related pesticides has been investigated more recently using soil mesocosm studies (Sanchez-Hernandez et al. 2017). β-Glucosidase activity was decreased by 43–59% after application of chlorpyrifos with two doses (4.8 and $24~kg~a.i.~ha^{-1}$). In the same study, β-glucosidase was found to respond differently in soil samples treated without and with sodium azide

(NaN$_3$), a potent inhibitor of microbial activity, as well as the type of pesticide. For instance, both chlorpyrifos and chlorpyrifos-oxon increased the activity of β-glucosidase in NaN$_3$-free soil suspensions compared with controls. In contrast, chlorpyrifos and 3,5,6-trichloro-2-pyridinol reduced β-glucosidase activity by 30% in NaN$_3$-soil suspension than controls. Furthermore, in the presence of NaN$_3$, β-glucosidase was highly sensitive to chlorpyrifos-oxon, wherein the activity was negatively affected even at lowest concentrations (10^{-8}–10^{-7} M) of the pesticide but not at higher lvels. This clearly indicates that the pesticide interaction with enzymes is mostly dose-dependent. However, inhibition of enzyme activity especially at lower concentration of pesticides might be due to exposure of binding sites to the enzyme.

β-Glucosidase has also been adversely affected by a herbicide diuron (Tejada et al. 2017). However, the enzyme responded differently to a commercial (Diuroey) and controlled release formulations (D-CRFs) of diuron in different soils. For instance, β-glucosidase activity decreased from 7 to 25 days after treatment with Diurokey and D-CRF in dystric Cambisol soil, and from 4 to 65 days after treatment with same herbicides in calcaric Fluvisol and LT soils. Nevertheless, enzyme activity was recovered to its initial value at the end of incubation period. More importantly, in a pure culture study involving an organophosphorous insecticide, selecron, at 10 and 50 ppm, the activities of C_x and C_1 cellulases of tested fungi were significantly decreased; however, C_1 cellulase of *A. fumigatus* was slightly increased (Omar et al. 1993). Likewise, using other pesticides that include a rice field herbicide, baythroid, Lodhi et al. (2000) observed an increase of 18.5% in cellulase activity at 16.5 μg g^{-1}, but the activity was declined by 25.9% at higher concentrations. Also, brominal and selecron inhibited cellulase activity in soil after most incubation periods (Omar and Abdel-Sater 2001).

Amendment with N-P-K fertilizers followed by prolonged incubation of soil samples not treated with an insecticide resulted in a significant increase in activity of cellulases (Fig. 4.2a). Activity of cellulases increased significantly (90–200%) after a single, two and three applications of acephate at 5.0 μg g^{-1} soil (Fig. 4.2a). But, at higher concentrations of acephate, the enzyme activities decreased by 10–29%, particularly after 2 and 3 repeated applications. The enzyme activity was more pronounced (29–80%) with buprofezin treatment, at 2.5–7.5 μg g^{-1} soil, together with fertilizer amendments. However, 10 μg g^{-1} soil treatment of the insecticide was slightly toxic to the enzyme. As with the soil samples that received no N-P-K fertilizers (Fig. 4.1), two repeated applications of the insecticides significantly enhanced the enzyme activity. The impact of buprofezin treatment to the soil coupled with N-P-K amendments was almost similar to that of acephate (Fig. 4.2b).

Interaction effects of the two selected insecticides, in combination, at graded concentrations of 0–10 μg g^{-1} soil, on the activities of soil cellulases were determined in unamended and NPK-amended soil samples. Acephate treatment, at 2.5 μg g^{-1} soil in combination with buprofezin up to 7.5 μg g^{-1} soil caused significant synergistic effect on the enzyme activity (Table 4.1). Additive response was observed only with 5 μg g^{-1} soil concentration of acephate and 2.5 μg g^{-1} soil of buprofezin. Higher concentrations above 5 μg g^{-1} soil of acephate or buprofezin

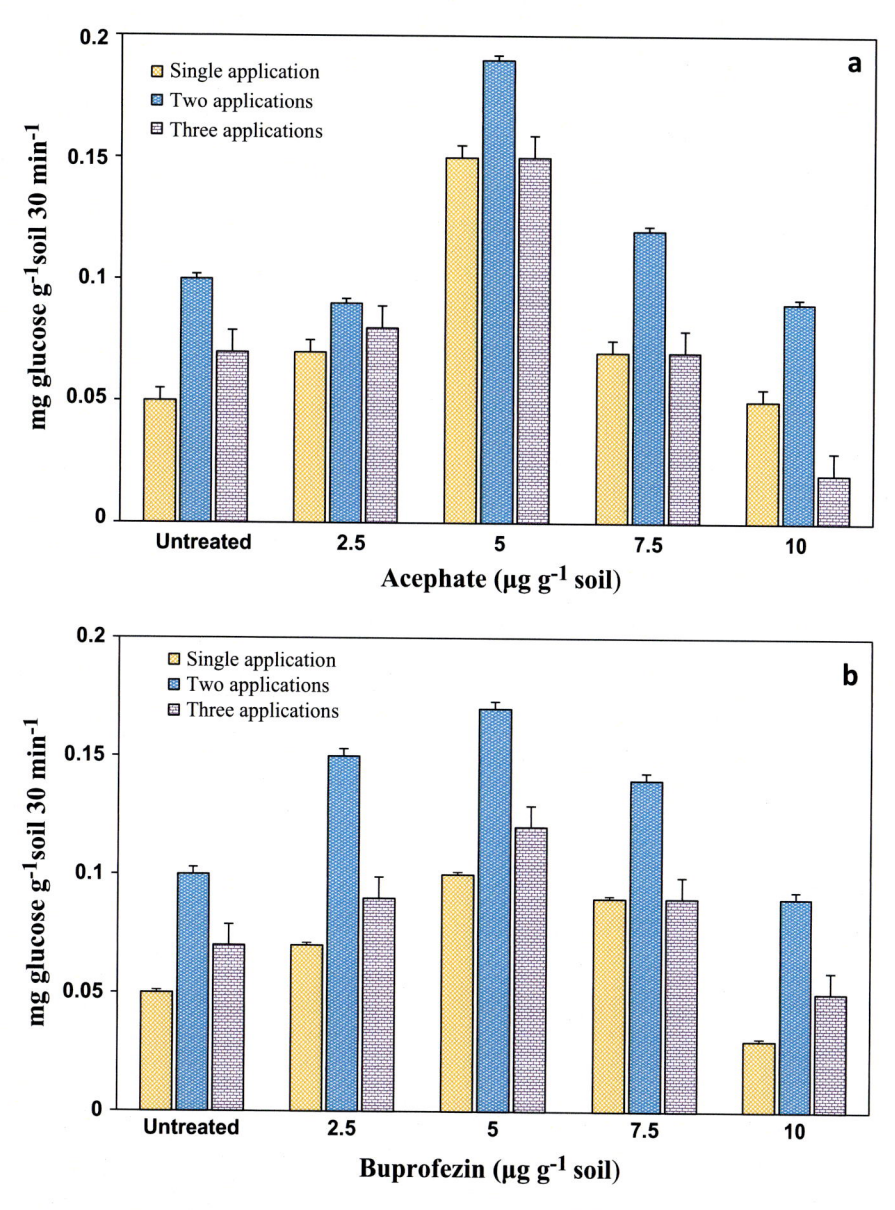

Fig. 4.2 Effect of single and repeated applications of (**a**) acephate and (**b**) buprofezin on activity of cellulases in soil amended with N-P-K. Error bars represent standard deviations ($n = 3$)

were significantly toxic resulting in antagonistic interaction towards the enzyme activity. Essentially, fertilizer amendments to the soil samples together with the insecticide combinations at graded levels had no measurable effect (Table 4.2) in altering the interaction effects noticed with the soil samples that received no N-P-K

Table 4.1 Interaction effects of insecticide combinations on activity of cellulases* in soil

		Buprofezin ($\mu g\ g^{-1}$ soil)				
		0	**2.5**	**5**	**7.5**	**10**
Acephate ($\mu g\ g^{-1}$ soil)	**0**	Control	0.07 ± 0.01 40[a]	0.12 ± 0.01 140	0.1 ± 0.05 100	0.02 ± 0.01 −60
	2.5	0.05 ± 0.03 1	0.16 ± 0.03[C] 220[a] 41[b]	0.24 ± 0.05[C] 380 140	0.17 ± 0.05[C] 240 100	0.55 ± 0.05[A] 1 −58
	5	0.17 ± 0.04 240	0.14 ± 0.02[B] 180 184	0.09 ± 0.02[C] 80 44	0.05 ± 0.02[A] 1 100	0.17 ± 0.06[A] 240 324
	7.5	0.04 ± 0.01 −20	0.16 ± 0.03[C] 220 28	0.17 ± 0.04[C] 240 148	0.07 ± 0.01[A] 40 100	0.05 ± 0.03[A] 1 92
	10	0.02 ± 0.03 −60	0.04 ± 0.01[A] −20 4	0.12 ± 0.05[A] 140 164	0.08 ± 0.03[A] 60 100	0.04 ± 0.01[A] −20 156

*mg glucose g^{-1} soil 30 min^{-1}
Control value, 0.05 ± 0.02 mg glucose g^{-1} soil 30 min^{-1}
All entries are means (n = 3) of per cent stimulation/inhibition values of enzyme activity relative to untreated control
[a]Experimental per cent values (first row) over control
[b]Expected per cent values (second row) over control
A: Significantly lower than the expected values, indicating an *antagonistic* insecticide interaction
B: Not significantly different from the expected values, indicating an *additive* insecticide interaction
C: Significantly greater than the expected values, indicating a *synergistic* insecticide interaction

fertilizers (Table 4.1). A similar observation was made by Gundi et al. (2007) that combination of monocrotophos and cypermethrin or quinalphos and cypermethrin, at different levels yielded synergistic and antagonistic responses towards cellulase activity in soil. These reports clearly indicate that there is no consistency on the impact of insecticide combinations towards cellulase activities in soil, one of the most useful parameters in toxicity studies. Again, this observation suggests that indiscriminate use of acephate and buprofezin, at higher rates but not at field applications rates, is deleterious to cellulases in soils.

Table 4.2 Interaction effects of insecticide combinations on activity of cellulases* in soil amended with N-P-K

		Buprofezin (µg g^{-1} soil)				
		0	**2.5**	**5**	**7.5**	**10**
Acephate (µg g^{-1} soil)	**0**	Control	0.07 ± 0.01 40[a]	0.1 ± 0.03 100	0.09 ± 0.01 80	0.05 ± 0.01 1
	2.5	0.07 ± 0.03 40	0.09 ± 0.01[C] 80[a] 64[b]	0.15 ± 0.04[C] 200 100	0.1 ± 0.02[C] 100 88	0.05 ± 0.02[A] 1 41
	5	0.15 ± 0.01 200	0.13 ± 0.02[B] 160 160	0.22 ± 0.04[C] 340 100	0.07 ± 0.03[A] 40 120	0.12 ± 0.01[A] 140 199
	7.5	0.07 ± 0.03 40	0.12 ± 0.02[C] 140 64	0.09 ±0.04[A] 80 100	0.07 ± 0.03[A] 40 80	0.05 ± 0.01[A] 1 41
	10	0.05 ± 0.02 1	0.04 ± 0.02[A] −20 41	0.07 ± 0.03[A] 40 100	0.03 ± 0.01[A] −40 80	0.04 ± 0.03[A] −20 2

*mg glucose g^{-1} soil 30 min^{-1}

Control value, 0.05 ± 0.02 mg glucose g^{-1} soil 30 min^{-1}

All entries are means (n = 3) of per cent stimulation/inhibition values of enzyme activity relative to untreated control

[a]Experimental per cent values (first row) over control

[b]Expected per cent values (second row) over control

A: *Antagonistic* insecticide interaction

B: *Additive* insecticide interaction

C: *Synergistic* insecticide interaction

References

Begum SFM, Rajesh G (2015) Impact of microbial diversity and soil enzymatic activity in dimethoate amended soils series of Tamil Nadu. Int J Environ Sci Technol 4:1089–1097

Endo T, Kuska T, Tan N, Sakai M (1982) Effects of the insecticide Cartap hydrochloride on soil enzyme activities, respiration and nitrification. J Pestic Sci 7:101–110

Gundi VAKB, Viswanath B, Chandra MS, Kumar VN, Reddy BR (2007) Activities of cellulase and amylase in soils as influenced by insecticide interactions. Ecotoxicol Envion Saf 68:278–285

Lodhi A, Malik NN, Mahmood T, Azam F (2000) Response of soil microflora, microbial biomass and some soil enzymes to Baythroid (an insecticide). Pak J Biol Sci 3:868–871

Nelson N (1944) A photometric adaptation of Somagyi method for determination of glucose. J Biol Chem 153:375–380

Omar SA, Abd-Alla MH (2000) Physiological aspects of fungi isolated from root nodules of faba bean (*Vicia faba* L.) Microbiol Res 154:339–347

Omar SA, Abdel-Sater MA (2001) Microbial populations and enzyme activities in soil treated with pesticides. Water Air Soil Pollut 127:49–63

Omar SA, Moharram AM, Abd-Alla MH (1993) Effects of an organophosphorus insecticide on the growth and cellulotytic activity of fungi. Int Biodeterior Biodegrad 31:305–310

Pancholy SK, Rice EL (1973) Soil enzymes in relation to old field succession: amylase, cellulase, invertase, dehydrogenase and urease. Soil Sci Soc Am Proc 37:47–50

Ross DJ, Speir TW (1985) Changes in biochemical activities of soil incubated with the nematicides and fenamiphos. Soil Biol Biochem 17:123–125

Ross DJ, Speir TW, Cowling JC, Whale KN (1984) Influence of field applications of oxamyl and fenamiphos on biochemical activities of soil under pasture. NZ J Sci 27:247–254

Sanchez-Hernandez JC, Sandoval M, Pierart A (2017) Short-term response of soil enzyme activities in a chlorpyrifos-treated mesocosm: use of enzyme-based indexes. Ecol Indic 73:525–535

Shittu OB, Akintokun AK, Akintokun PO, Gbadebo MO (2004) Effect of diazinon application on soil properties and soil microflora. Proceedings of International Conference and National Development, 25–28 October. COLNAS Proceedings, pp 68–74

Srinivasulu M, Rangaswamy V (2006) Activities of invertase and cellulase as influenced by the application of tridemorph and captan to ground nut (*Araachis hypogeae*) soil. Afr J Biotechnol 5:175–180

Tejada M, Morillo E, Gómez I, Madrid F, Undabeytia T (2017) Effect of controlled release formulations of diuron and alachlorherbicides on the biochemical activity of agricultural soils. J Hazard Mater 322:334–347

Chapter 5
Impact of Acephate and Buprofezin on Soil Amylases

Assay of Amylases in Soil

Total amylase activity in untreated and insecticide- and/or fertilizer-treated soil samples was determined following the method developed by Cole (1977) and modified by Tu (1981a, b). Five grams of soil samples were placed in test tubes (25 × 200 mm), and 1.0 mL of toluene was added. The contents in the tubes were mixed thoroughly, and 6 mL of 2% starch in 0.2 M acetate buffer (pH 5.5) was added after 15 min. After incubating the tubes for 48 h, the soil suspension was passed through Whatman No.1 filter paper. The amount of reducing sugar content in the filtrate was determined by Nelson-Somagyi method in a digital spectrophotometer (Elico). Glucose was used as a standard. Amylase activity was expressed in terms of mg glucose g^{-1} soil 48 h^{-1}.

Nontarget Effects on Amylases

Soil amylases, one of the important indices of microbial activity, were assayed after single or repeated applications of acephate and buprofezin at and above the field rates applied in the present study. The activity was measured in terms of glucose accumulated from starch. Amylase activity was significantly stimulated (5–58%) by a single application of acephate up to 10.0 $\mu g\ g^{-1}$ soil over untreated control (Fig. 5.1a). However, most pronounced effect was noticed at 5 $\mu g\ g^{-1}$ soils, and there was a deceasing (18–24%) tendency in enzyme activity afterwards. After two and three repeated applications of acephate, there was a progressive decrease in the accumulation of glucose. The concentration of buprofezin that effected a significant stimulation (5–67%) in enzyme activity after single application in soil samples ranged from 2.5 to 7.5 $\mu g\ g^{-1}$. The accumulation of glucose was more striking at 7.5 $\mu g\ g^{-1}$. Nevertheless, profound negative influence of the enzyme activity was noticed at 10 $\mu g\ g^{-1}$ concentration (Fig. 5.1b).

© Springer International Publishing AG 2018
N.R. Maddela, K. Venkateswarlu, *Insecticides–Soil Microbiota Interactions*,
DOI 10.1007/978-3-319-66589-4_5

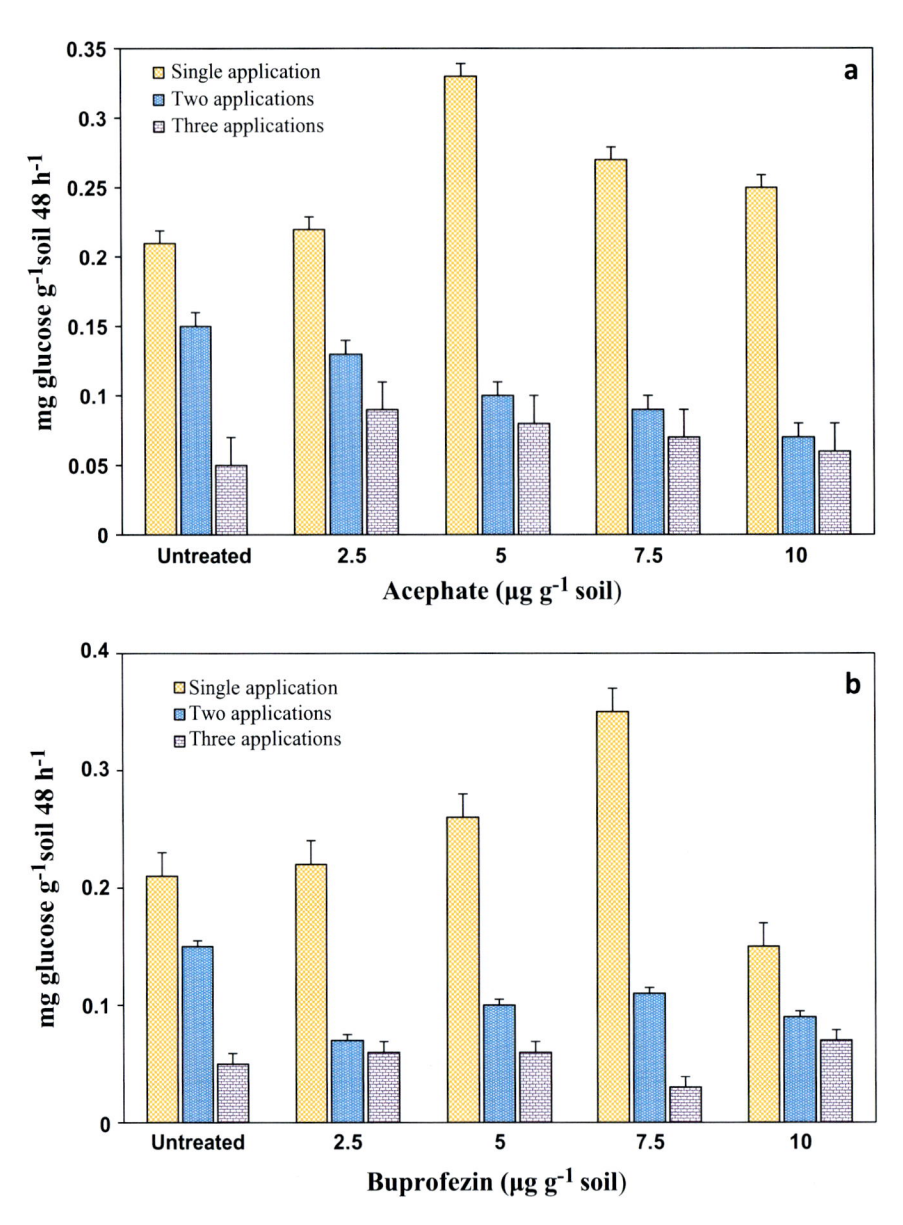

Fig. 5.1 Effect of single and repeated applications of (**a**) acephate and (**b**) buprofezin on activity of amylases in soil. Error bars represent standard deviations (n = 3)

As with acephate, repeated applications (two or three) of buprofezin to soil adversely affected the amylase activity. A similar trend was also noticed with the applications of either acephate or buprofezin in N-P-K fertilizer-amended soil (Fig. 5.2a and b). Application of buprofezin up to 7.5 μg g^{-1} greatly enhanced (60–190%) the activity of

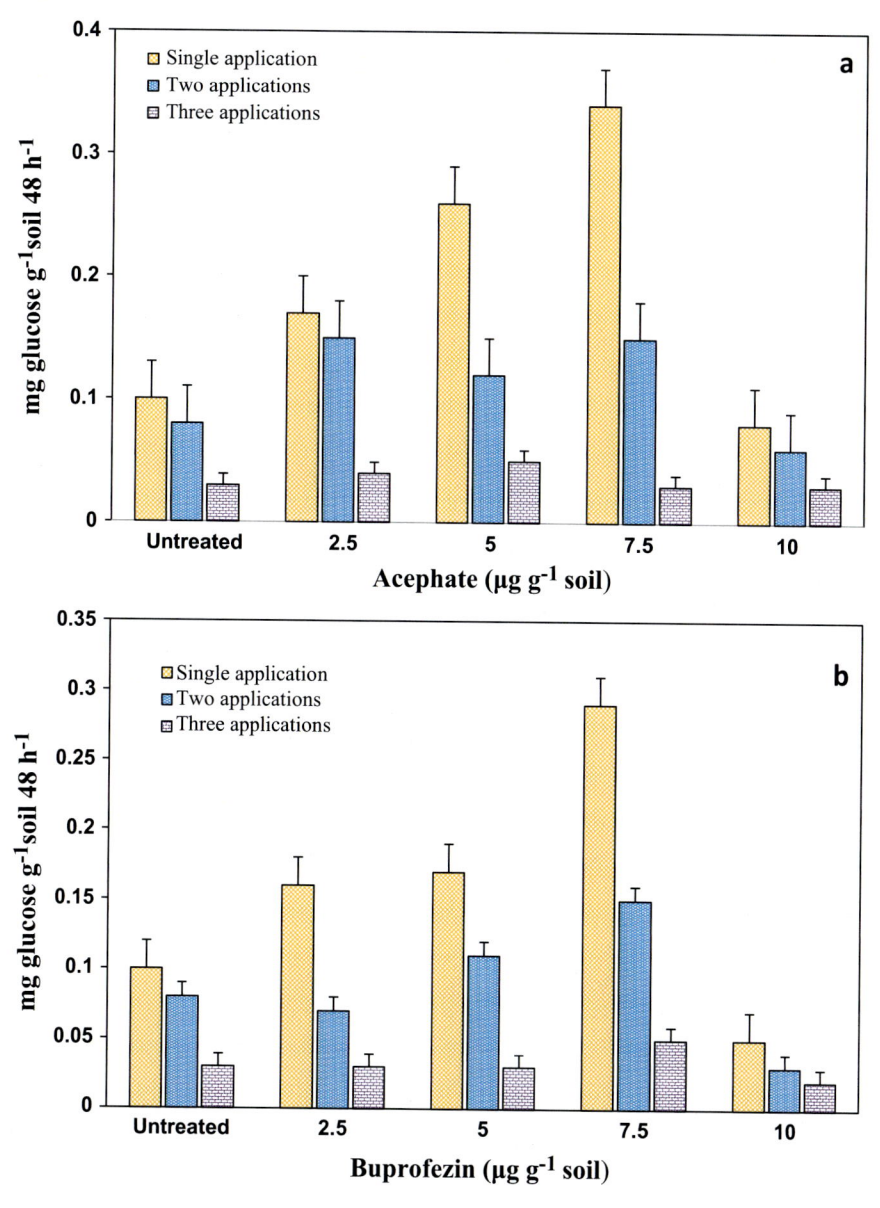

Fig. 5.2 Effect of single and repeated applications of (**a**) acephate and (**b**) buprofezin on activity of amylases in soil amended with N-P-K. Error bars represent standard deviations (n = 3)

amylases in NPK-amended soils. But, this toxicity parameter was negatively affected at 10 µg g^{-1} soil (Fig. 5.2b). Though single application of buprofezin stimulated the activity of amylases in soil amended with nutrient fertilizer, two or three repeated applications resulted in inhibition of the enzyme activity. Thus, there was no significant difference on the effect of amylases as influenced by single or repeated applications of acephate and buprofezin.

In fact, amylases are likely to be affected by addition of pesticides as observed in several studies (Rangaswamy and Venkateswarlu 1992; Tu 1995; Mandic et al. 1997; Ismail et al. 1998; Lodhi et al. 2000; Singh et al. 2002; and Gundi et al. 2007). Endosulfan at 32 and 48 μL L^{-1} was reported to stimulate α-amylase activity in the supernatant of culture filtrate, but the enzyme activity was adversely affected at 64 and 80 μL L^{-1} (Tolan and Ensari 2006). Similarly, activity of amylases was enhanced by monocrotophos, quinalphos, cypermethrin or fenvalerate at levels ranging from 1.0 to 2.5 kg ha^{-1}, but was inhibited at concentrations of 5 and 10 kg ha^{-1} in ground-nut soils (Rangaswamy and Venkateswarlu 1992). Gundi et al. (2007) reported that monocrotophos, quinalphos or cypermethrin effected an increase in amylase activity in black vertisol and red alfinsol soil at all three concentrations (5, 10 and 25 μg g^{-1} soil) after 10 days of incubation. Several pesticides including organophosphates at 5 and 10 mg kg^{-1} were found to enhance amylase activity (Tu 1982). Also, Tu (1988) reported that malathion and permethrin at higher levels were stimulatory to amylase activity 3 days after the application. But, no effects were observed with tefluthrin, DOWCO 429X and DPX 43898 at 10 mg kg^{-1} in a sandy loam soil (Tu 1990). On the contrary, application of cyfluthrin and imidacloprid to soil initially inhibited amylase activity followed by the stimulatory effect at the end of 3 weeks of incubation (Tu 1995). Also, at a concentration of 930 mg kg^{-1}, fenamiphos had a deleterious effect on amylase activity, which was reduced by 24% after 62 days of treatment under laboratory conditions (Ross and Speir 1985).

The changes of soil amylase activity in response to simultaneous and sequential applications of several pesticides were studied in laboratory experiments. Lodhi et al. (2000) observed an increase in amylase activity by a maximum of 91.5% at 1.6 μg g^{-1} of baythroid. At higher concentration of 6.4 μg baythroid g^{-1} soil, however, the activity was reduced by 47.9%. Similarly, metsulfuron-methyl at 5 μg g^{-1} caused a reduction in amylase activity (Ismail et al. 1998). Tu (1995) found inhibitory effect of imidacloprid on amylase activity after a week, while significant recovery was observed after three wks. Also, amylase activity was affected when soil treated with dimethoate (Mandic et al. 1997) or chlorothalonil (Singh et al. 2002). In contrast, application of monocrotophos, quinalphos or cypermethrin greatly enhanced the activity of amylases in soil (Gundi et al. 2007).

Even in the recent past, activities of amylases in soil were assessed against several kinds of insecticides and fungicides. For example, soil treatment with 1% dimethoate caused significant adverse effects on amylase activity in three different soils (Begum and Rajesh 2015). During the incubation period (0–120 days), amylase activity was higher in untreated controls than that in treated samples. However, with fewer exceptions, the trend in the response of amylase was very similar among the samples in the period of study. Enzyme activity was 50% higher during 30–60 days incubation, but thereafter, the activity decreased gradually and reached to the original at day 120. Ten days after treatment, two insecticides, thiodicarb and dimethoate, significantly enhanced the activity of amylase at lower concentrations (1.0–5.0 kg ha^{-1}), but the enzyme activity was adversely affected at higher doses of 7.5–10.0 kg ha^{-1} (Rekhapadmini et al. 2016). Lindane (gamma-hexachlorocyclohexane, γ-HCH) was tested at concentrations of 1.0, 2.0 and 3.0 kg a.i. ha^{-1} on amylolytic bacteria and fungi, and pure cultures of soil bacteria by Zargar and Johri (1995), and observed an increase in bacterial population with increasing incubation period until 18 days, there after, a decline. The highest concentration of 3.0 kg ha^{-1} was more toxic to amylolytic bacteria than that of other two concentrations tested. On the other hand, amylolytic fungi were more sensitive to γ-HCH than bacteria, but amylolytic fungi recovered later. However, the bacterial population slightly increased after 45 days, probably due to the degradation of γ-HCH, and thus releasing stress on amylolytic bacteria (Reed and Forgash 1969), and the recovery of amylolytic fungi after 18 days might be due to the isomerization of γ-HCH to its α-form (Sahu et al. 1990). Pure culture of a soil isolate, *Bacillus* sp., showed 1.2 h of lag phase, 4.5 h of logarithmic phase followed by stationary phase in medium supplemented with starch (1%) that contained γ-HCH (5.0, 10.0 and 15.0 mg L^{-1}) (Zargar and Johri 1995). Even though 5.0 mg L^{-1} of γ-HCH showed marginal effects on growth of *Bacillus* sp., higher rates (10.0 and 15.0 mg L^{-1}) of the insecticide significantly affected its growth. Response of amylase to fungicide was also similar to that of insecticides. Likewise, the activity of amylase increased at 10 ppm of mancozeb, but the enzyme activity decreased significantly with increasing concentrations (10 to 100 ppm) of mancozeb (Walia et al. 2014). Furthermore, amylase activity decreased after 2 week of incubation; however, the activity was restored to the original level (observed on 0-day) after three- and four-weeks of incubation (Walia et al. 2014). Response of amylases in soil treated with more than one chemical is also similar to that of soil treated with a single chemical. For instance, soil application of imidacloprid (insecticide) and triadimefon (fungicide) alone or in combination resulted in significant decrease of amylase activity (Deborah et al. 2013). At field rate (0.5 μg g^{-1}), the pesticides stimulated the activity of amylase at day 10. However, the enzyme activity declined significantly at the higher doses of 0.7 μg g^{-1}.

A differential response of amylases was observed in soils with and without NPK-fertilizer amendments when acephate and buprofezin were present at graded concentrations in the mixture. Acephate at a rate of 2.5 μg g^{-1} together with buprofezin up to 7.5 μg g^{-1} exerted a synergistic response on activity of amylases (Table 5.1). However, the combination at 5 μg g^{-1} showed an additive effect on enzyme activity. Nonetheless, higher concentrations of the two insecticides, in combination, resulted in significant antagonistic effect on soil amylases.

On the other hand, combination of the two insecticides interacted more antagonistically with amylases in soil that received NPK-fertilizer (Table 5.2). Especially, acephate treatment, at 7.5 and 10 μg g^{-1} soils in combination with buprofezin from 2.5 to 10 μg g^{-1} soil caused exclusively antagonistic effect on activity of amylases. Thus, the interaction effects of both the insecticides were more adverse to the enzyme in NPK-amended soil when compared with the soil that did not receive the

Table 5.1 Interaction effects of insecticide combinations on activity of amylases* in soil

		Buprofezin (μg g^{-1} soil)				
		0	**2.5**	**5**	**7.5**	**10**
Acephate (μg g^{-1} soil)	**0**	Control	0.22 ± 0.09 5[a]	0.26 ± 0.03 24	0.35 ± 0.09 67	0.15 ± 0.03 −29
	2.5	0.22 ± 0.04 5	0.40 ± 0.06[C] 90[a] 9[b]	0.73 ± 0.22[C] 248 27	0.38 ± 0.08[C] 81 68	0.25 ± 0.05[A] 19 −22
	5	0.33 ± 0.09 57	0.35 ± 0.03[C] 67 59	0.34 ± 0.01[B] 62 67	0.31 ± 0.16[A] 48 86	0.25 ± 0.05[A] 19 45
	7.5	0.27 ± 0.01 29	0.41 ± 0.06[C] 95 32	0.27 ± 0.03[A] 29 46	0.30 ± 0.09[A] 43 76	0.21 ± 0.03[A] 1 8
	10	0.25 ± 0.10 19	0.25 ± 0.05[A] 19 23	0.28 ± 0.07[A] 33 38	0.24 ± 0.06[A] 14 70	0.28 ± 0.04[A] 33 −4

*mg glucose g^{-1} soil 48 h^{-1}

Control value, 0.21 ± 0.08 mg glucose g^{-1} soil 48 h^{-1}

All entries are means (n = 3) of per cent stimulation/inhibition values of enzyme activity relative to untreated control

[a]Experimental per cent values (first row) over control

[b]Expected per cent values (second row) over control

A: *Antagonistic* insecticide interaction

B: *Additive* insecticide interaction

C: *Synergistic* insecticide interaction

nutrients. Gundi et al. (2007) reported that an increment of about 51% of amylase activity occurred with the combination of monocrotophos and cypermethrin at 5 μg g^{-1} by the end of 10 days incubation, but the increments observed when treated with the insecticides alone were 30% and 17%, respectively. The present results indicate that higher rates of insecticide application, either one or more, and in combination greatly affect soil amylases.

Table 5.2 Interaction effects of insecticide combinations on activity of amylases* in soil amended with N-P-K

		Buprofezin (μg g^{-1} soil)				
		0	**2.5**	**5**	**7.5**	**10**
Acephate (μg g^{-1} soil)	**0**	Control	0.16 ± 0.06 / 60[a]	0.17 ± 0.05 / 70	0.29 ± 0.08 / 190	0.13 ± 0.05 / 30
	2.5	0.17 ± 0.04 / 70	0.24 ± 0.06[C] / 140[a] / 88[b]	0.24 ± 0.05[C] / 140 / 91	0.29 ± 0.07[C] / 190 / 127	0.14 ± 0.06[A] / 40 / 79
	5	0.26 ± 0.15 / 160	0.37 ± 0.03[C] / 27 / 124	0.42 ± 0.05[C] / 320 / 118	0.31 ± 0.03[C] / 210 / 46	0.22 ± 0.07[A] / 120 / 142
	7.5	0.34 ± 0.05 / 240	0.22 ± 0.06[A] / 120 / 156	0.21 ± 0.08[A] / 110 / 142	0.31 ± 0.05[A] / 210 / −26	0.21 ± 0.23[A] / 110 / 198
	10	0.32 ± 0.06 / 220	0.15 ± 0.08[A] / 50 / 124	0.28 ± 0.16[A] / 180 / 236	0.22 ± 0.14[A] / 120 / −8	0.24 ± 0.04[A] / 140 / 184

*mg glucose g^{-1} soil 48 h^{-1}

Control value, 0.10 ± 0.04 mg glucose g^{-1} soil 48 h^{-1}

All entries are means (n = 3) of per cent stimulation/inhibition values of enzyme activity relative to untreated control

[a]Experimental per cent values (first row) over control

[b]Expected per cent values (second row) over control

A: *Antagonistic* insecticide interaction

B: *Additive* insecticide interaction

C: *Synergistic* insecticide interaction

References

Begum SFM, Rajesh G (2015) Impact of microbial diversity and soil enzymatic activity in dimethoate amended soils series of Tamil Nadu. Int J Environ Sci Technol 4:1089–1097

Cole MA (1977) Lead inhibition of enzyme synthesis in soil. Appl Enviorn Microbiol 33:262–268

Deborah BV, Mohiddin MJ, Madhuri RJ (2013) Interaction effects of selected pesticides on soil enzymes. Toxicol Int 20:195–200

Gundi VAKB, Viswanath B, Chandra MS, Kumar VN, Reddy BR (2007) Activities of cellulase and amylase in soils as influenced by insecticide interactions. Ecotoxicol Enviorn Saf 68:278–285

Ismail BS, Yapp KF, Omar O (1998) Effects of metsulfuron-methyl on amylase, urease and protease activities in two soils. Aust J Soil Res 36:449–456

Lodhi A, Malik NN, Mahmood T, Azam F (2000) Response of soil microflora, microbial biomass and some soil enzymes to Baythroid (an insecticide). Pak J Biol Sci 3:868–871

Mandic L, Dukic D, Govedarica M, Kovic SS (1997) The effect of some insecticides on the number of amylolytic microorganisms and *Azotobacter* in apple nursery soil. Czeckoslevensko Vocarstvo 31:177–184

Rangaswamy V, Venkateswarlu K (1992) Activities of amylase and invertase as influenced by the applications of monocrotophos, quinalphos, cypermethrin and fenvalerate to groundnut soil. Chemosphere 25:525–530

Reed WT, Forgash AJ (1969) Metabolism of lindane to tetrachlorobenzene. J Agric Food Chem 17:896–897

Rekhapadmini A, Anuradha B, Rangaswamy V (2016) Impact of insecticides thiodicarb and dimethoate on soil microbial activities (amylase) in two groundnut (*Arachis hypogaea* L.) soils. Int J Recent Sci Res 7:9764–9768

Ross DJ, Speir TW (1985) Changes in biochemical activities of soil incubated with the nematicides and fenamiphos. Soil Biol Biochem 17:123–125

Sahu SK, Patnaik KK, Sethunathan N (1990) Degradation of α-, β- and γ-isomers of hexachlorocyclohexane by rhizosphere soil suspension from sugarcane. Proc Indian Acad Sci (Plant Sci) 100:165–172

Singh BK, Allan W, Denus JW (2002) Degradation of chlorpyrifos, fenamiphos and chlorothalonil alone and in combination and their effects on soil microbial activity. Environ Toxicol Chem 21:2600–2605

Tolan V, Ensari Y (2006) Effect of endosulfan on growth, α-amylase activity and plasmid amplification in *Bacillus subtilis*. Indian J Biochem Biophys 43:123–126

Tu CM (1981a) Effect of pesticides on activity of enzymes and microorganisms in a clay loam soil. J Environ Sci Health B16:179–191

Tu CM (1981b) Effect of some pesticides on enzyme activities in an organic soil. Bull Environ Contam Toxicol 27:109–114

Tu CM (1982) Influence of pesticides on activities of amylase, invertase and level of adenosine triphosphate in organic soil. Chemosphere 2:909–914

Tu CM (1988) Effect of selected pesticides on activities of invertase, amylase and microbial respiration in sandy soil. Chemosphere 17:159–163

Tu CM (1990) Effects of four experimental insecticides on enzyme activities and levels of adenosine triphosphate in mineral and organic soils. J Environ Sci Health B25:787–800

Tu CM (1995) Effect of five insecticides on microbial and enzymatic activities in sandy soil. J Environ Sci Health B30:289–306

Walia A, Mehta P, Guleria S, Chauhan A, Shirkot CK (2014) Impact of fungicide mancozeb at different application rates on soil microbial populations, soil biological processes, and enzyme activities in soil. Sci World J 2014:1–9

Zargar MY, Johri BN (1995) Effect of gamma-hexachlorocyclohexane on amylolytic microorganisms of soil and amylase activity. Bull Environ Contam Toxicol 55:426–430

Chapter 6
Impact of Acephate and Buprofezin on Soil Invertase

Assay of Invertase in Soil

Invertase activity of the test soil samples was determined following the method of Tu (1982). Five grams of soil samples were placed in test tubes (25 × 200 mm), and 1.0 mL of toluene was added. The contents in the tubes were mixed thoroughly. After 15 min, 6 mL of 18 mM sucrose in 0.2 M acetate buffer (pH 5.5) was added, and the tubes were incubated for 6 h. The suspension was filtered through Whatman No.1 paper, and the amount of reducing sugar in the filtrate was determined by Nelson-Somagyi method in a digital spectrophotometer (Elico). Glucose was used as a standard. The enzyme activity was expressed as milligrams of glucose released per g of soil in 6 h (mg glucose g^{-1} soil 6 h^{-1}).

Nontarget Effects on Invertase

Impact of acephate and buprofezin on soil invertase activity was investigated after a single or repeated application in concentrations ranging from 0 to 10 $\mu g\ g^{-1}$. Single application of acephate at lower concentrations (2.5–7.5 $\mu g\ g^{-1}$) caused stimulation (73–209%) to invertase activity in soil, but was toxic to the enzyme at higher concentrations (Fig. 6.1a). In particular, there was a pronounced stimulation in the activity of invertase at 5 $\mu g\ g^{-1}$ which is close to the field application rate of acephate. Also, two or three repeated applications of acephate could adversely affect the invertase activity in soil. Similar to the above findings, single application of buprofezin, up to 7.5 $\mu g\ g^{-1}$, was stimulatory to the enzyme activity. The increase in the activity was maximal (264%) at 5 μg buprofezin g^{-1} soil. However, the highest level (10 $\mu g\ g^{-1}$) of buprofezin significantly inhibited (45%) the invertase activity (Fig. 6.1b). Furthermore, two or three repeated applications of buprofezin caused adverse effects to invertase in soil. The enzyme activity was stimulated after a single addition, but there was a progressive decline after three repeated applications.

© Springer International Publishing AG 2018
N.R. Maddela, K. Venkateswarlu, *Insecticides–Soil Microbiota Interactions*,
DOI 10.1007/978-3-319-66589-4_6

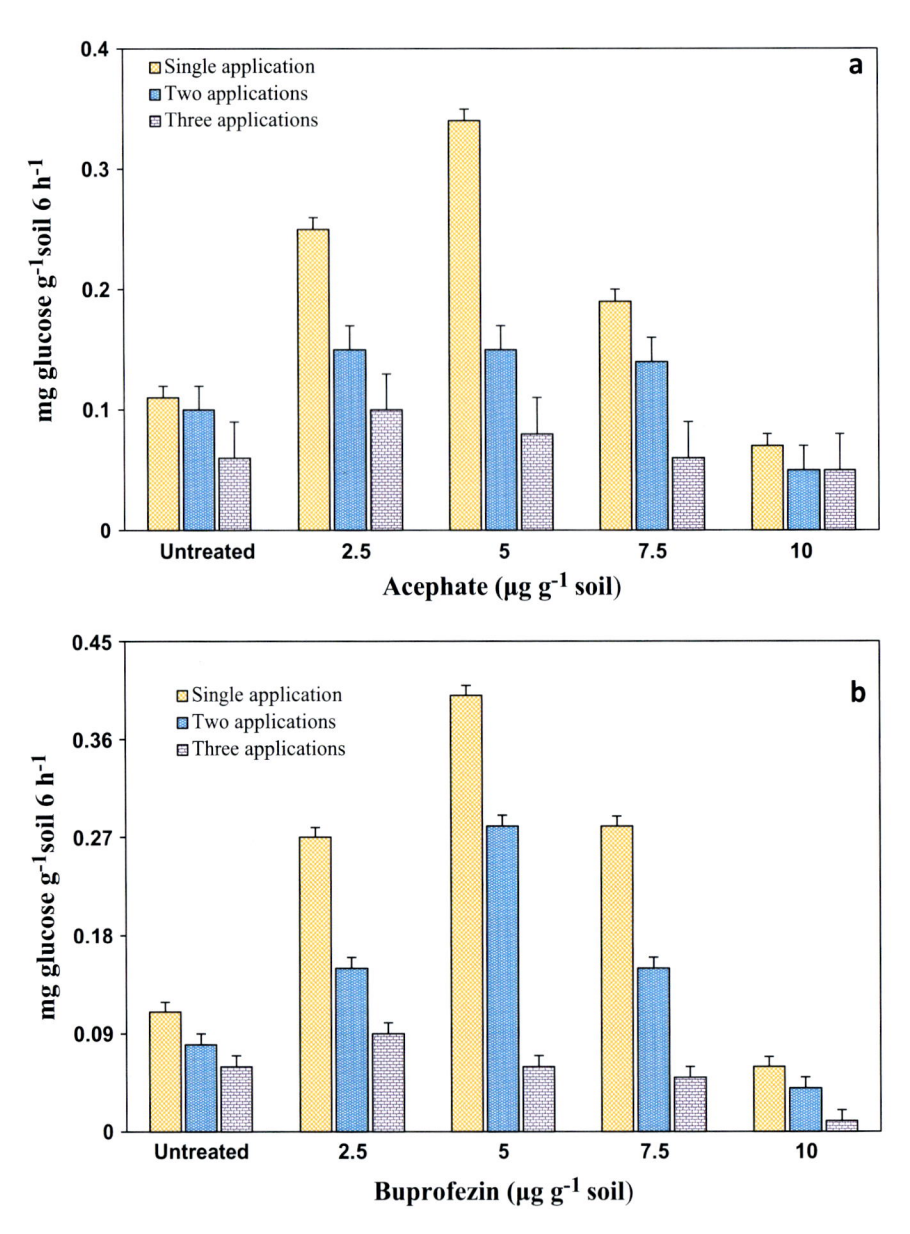

Fig. 6.1 Effect of single and repeated applications of (**a**) acephate and (**b**) buprofezin on invertase activity in soil. Error bars represent standard deviations (n = 3)

Although several anthropogenic chemicals have been assessed for their nontarget effects toward soil invertase, organophosphorous insecticides have received very less attention. For instance, invertase activity increased by 110.9% at 1.6 μg baythroid g^{-1} soil and decreased by 40.3% at the highest level studied (Lodhi et al. 2000). Srinivasulu and Rangaswamy (2006) reported an increase in invertase activity in soil with increasing concentrations of fungicides; the enzyme activities were adversely affected at higher rates of each test chemical. Similarly, carbaryl, a methylcarbamate insecticide, when applied at a normal agricultural dose did not have any inhibitory effect on invertase activity in soil (Mishra and Pradhan 1987). In contrast, Tu (1993) suggested that captafol and chlorothalonil suppressed invertase activity for one day temporarily in a sandy loam soil and later on, after 2 days, the inhibitory effect was alleviated. Similar observation of alleviation was also observed with dimethoate (Begum and Rajesh 2015). One per cent of dimethoate adversely affected the activity of invertase in three different soils in the first 45 days of incubation. In fact, ~ 30% of the enzyme activity decreased in soils that received 1% dimethoate than control soils during 0–45 days of incubation. Subsequently, from day 60 to 120 about 10–40% higher activity was recorded in dimethoate-treated soils than controls. Voets et al. (1974) also showed that long-term atrazine applications significantly reduced the activity of invertase in soil. Higher enzyme activity in test samples over control at later stages of incubation might be due to the utilization of pesticide by adapted soil microflora or reduced toxicity of pesticide due to its physico-chemical interactions with soil particles.

A study was conducted with chlorothalonil (Yu et al. 2006) to evaluate its effects on invertase activity in soil after repeated applications. After the first addition, activity was significantly reduced, but marked inhibition was observed after second application. The most important impact of these findings was the transient negative effects that became weaker following the third and fourth treatments. Similarly, soil treated with monosulfuron (Yong-hong and Yu-bao 2005), carbaryl and atrazine (Gianfreda and Sannino 1993) resulted in an obvious inhibition of invertase activity in soil. Atrazine and simaizne affected invertase activity adversely in the beginning, but activity was stimulated in the later stages (Schäffer 1993). Surprisingly, soil invertase activity was not sensitive to omethoate (Xiang et al. 2009), oxamyl, enamiphos, and 2,4-D (Schäffer 1993), and other insecticides such as pyrethirns and Neemix-4E (Antonions 2003). In contrast, Lodhi et al. (2000) reported that invertase activity increased by 110.9% at 1.6 µg baythroid g^{-1} soil followed by a decrease of 40.3% at the highest level tested (6.4 µg g^{-1} soil). Also, glyphosate and paraquat increased soil invertase activity (Gianfreda and Sannino 1993; Sannino and Gianfreda 2001).

Several fungicides were also tested against the activity of soil invertase in the recent past. With the application of a fungicide, mancozeb, at concentrations up to 10 ppm, soil invertase activity increased by 11% (Walia et al. 2014). There were no significant changes in the activity of invertase at higher doses (100, 250, 500 and 100 ppm) of this fungicide. Furthermore, incubation time showed varying effect on activity of invertase at almost all the concentrations of mancozeb. In a very recent study, Wang et al. (2017) found that the activity of invertase was stimulated by a fungicide, dimethomorph, at all the concentrations (1, 10, and 100 mg kg^{-1}) throughout the incubation period of 28 days. In another study, invertase activity was stimulated by individual and combination of fungicides, tebuconazole (TEB) and carbendazim (CAB), at concentrations of 1.0 and 10 mg kg^{-1} (Wang et al. 2016). Stimulation was continued until day 30, but without any exception, the enzyme activity was adversely affected throughout 90-day incubation period. A novel fungicide, pyrimorph, significantly influenced the activities of invertase (Xiong et al. 2013). Even though no differences were observed in the activities of invertase at the applied doses (0.5 and 5 mg kg^{-1}) during the first 60 days, 30–73% increment in enzyme activity was recorded at 90th day. Furthermore, higher doses of pyrimorph (50, 100 and 150 mg kg^{-1}) increased the enzyme activities slightly at day 7 and clearly at day 14. Thus, the enzyme activities (per cent increment) were 121 and 184% after 14 days in soils that received pyrimorph at 50 and 150 mg kg^{-1}, respectively. Nevertheless, there was a remarkable reduction in the invertase activity after 21 days. Application of two fungicides, imidacloprid and triadimefon, to soil at 0.2, 0.5, 0.7 µg g^{-1} either singly or in combination caused adverse effects on invertase activity (Deborah et al. 2013). On the contrary, increase in invertase activity was 13.6% on day 4 in soil received that received 4.0 mg kg^{-1} carbendazim (fungicide), but the acitivities decreased by 10.2% and 15.3% on days 23 and 30, respectively (Yan et al. 2011). Carbendazim concentration at 8.0 mg kg^{-1} was inhibitory even on day 1. Furthermore, invertase response to a combination of carbendazim (4.0 mg kg^{-1}) and chloramphenicol (10.0 and 20.0 mg kg^{-1}) was very similar to that of soil received with carbendazim alone (Yan et al. 2011). The enzyme was more adversely affected when the fungicide-treated soil was incubated with a substrate, sucrose, for 24 h.

The data presented in Fig. 6.2 indicate that the activity of invertase was much more affected in NPK-fertilizer amended soils upon single or repeated applications of the two insecticides over those soil samples with no fertilizer amendments. After single application of acephate, significant stimulation (44%) of the enzyme activity was noticed at 5 µg g^{-1}; however, the activity was negatively affected at higher rates of acephate (Fig. 6.2a). Furthermore, greater accumulation of glucose was noticed after single application of acephate to NPK-amended soil; the activity was greatly declined later. Indeed, the activity was decreased to 50% after two or three repeated applications over the activity observed after a single application. On the other hand, response of invertase to single or repeated applications of buprofezin in amended soils was almost parallel to that of acephate treatment (Fig. 6.2b).

All the three different interaction responses emerged with the combinations of acephate and buprofezin towards the activity of invertase in soil (Table 6.1). Particularly, acephate and buprofezin at equal concentrations of 2.5 µg g^{-1} in combination caused synergistic stimulation on soil invertase activity. Additive response was observed only with 5 µg g^{-1} soil of acephate and 2.5 µg g^{-1} soil of buprofezin. However, combinations of 7.5 or 10 µg g^{-1} acephate with graded concentrations of buprofezin elicited antagonistic response towards invertase activity in soil. Similarly, in NPK-amended soils, the two selected insecticides in combinations have exerted exclusively synergistic stimulations and antagonistic inhibitions on invertase activities at lower and higher rates, respectively (Table 6.2). Equal concentrations of the two insecticides in combination at 2.5 or 5 µg g^{-1} soils have synergistically stimulated the invertase activity, whereas 7.5 or 10 µg g^{-1} soil could inhibit the activity antagonistically. From the above data, it can be concluded that invertase is highly sensitive to the selected insecticides when applied alone or in combinations. But, amendment of soils with N-P-K fertilizers seems to alleviate the toxicity of the insecticides towards invertase. Bielinska and Pranagal (2007)

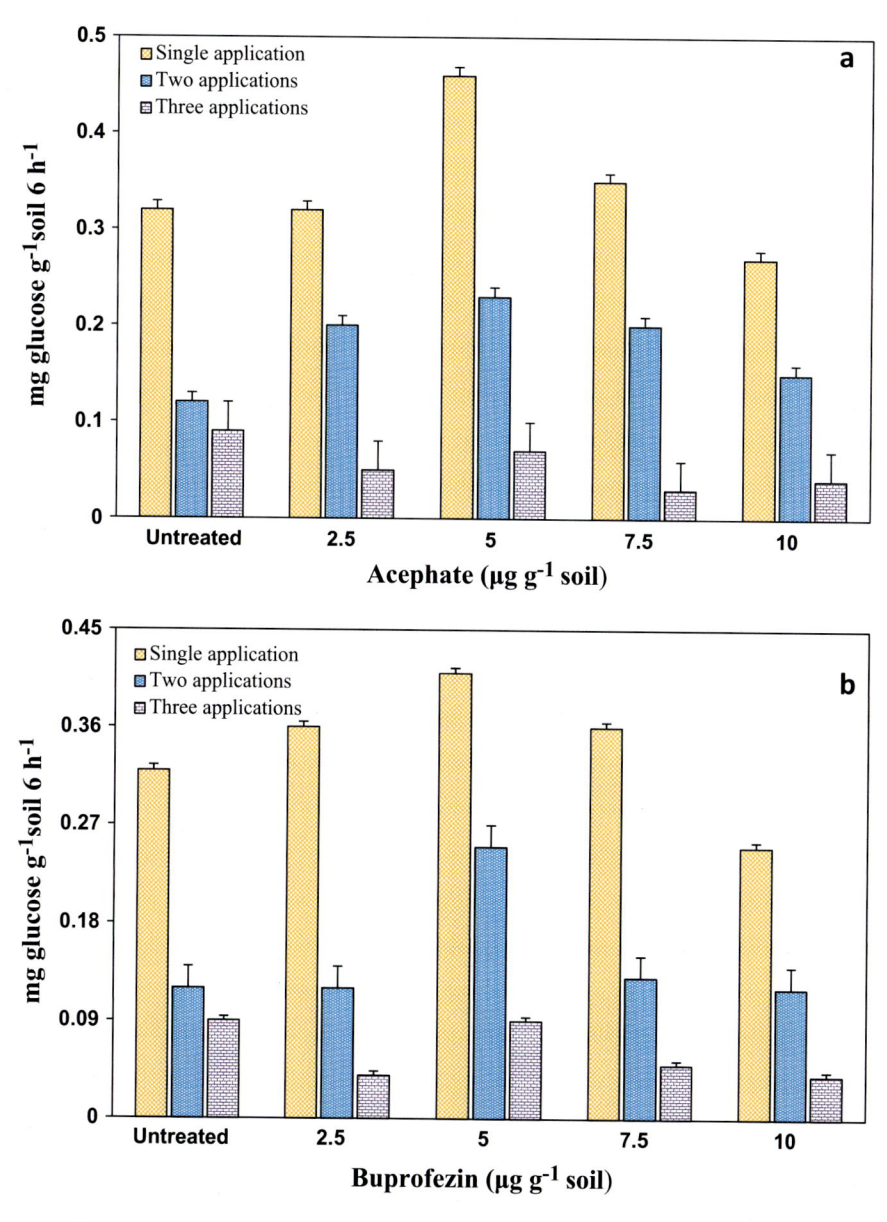

Fig. 6.2 Effect of single and repeated applications of (**a**) acephate and (**b**) buprofezin on invertase activity in soil amended with N-P-K. Error bars represent standard deviations (n = 3)

showed that the application of high level of mineral fertilization simultaneously with chemical weed control appears to be detrimental particularly to biological activity of the soil. They also suggested that the accompanying decrease in pH was an additional cause of the strong lowering in soil enzymatic activity for that site.

Table 6.1 Interaction effects of insecticide combinations on invertase activity* in soil

		Buprofezin (μg g^{-1} soil)				
		0	**2.5**	**5**	**7.5**	**10**
Acephate (μg g^{-1} soil)	**0**	Control	0.27 ± 0.01 145[a]	0.64 ± 0.13 482	0.30 ± 0.10 173	0.20 ± 0.01 82
	2.5	0.25 ± 0.10 127	0.68 ± 0.01[C] 518[a] 203[b]	1.94 ± 0.01[C] 1664 967	0.24 ± 0.02[A] 118 265	0.22 ± 0.06[A] 100 105
	5	0.34 ± 0.11 209	0.37 ± 0.12[B] 236 240	1.6 ± 0.13[C] 1354 1280	0.11 ± 0.03[A] 1.0 325	0.02 ± 0.01[A] −82 120
	7.5	0.19 ± 0.03 73	0.29 ± 0.01[C] 164 112	0.10 ± 0.02[A] −9 203	0.06 ± 0.02[A] −45 120	0.05 ± 0.01[A] −55 95
	10	0.07 ± 0.01 −36	0.16 ± 0.04[A] 45 161	0.05 ± 0.04[A] −54 619	0.11 ± 0.02[A] 1 235	0.08 ± 0.01[A] −27 75

* mg glucose g^{-1} 6 h^{-1}

Control value, 1.11 ± 0.08 mg glucose g^{-1} 6 h^{-1}

All entries are means (n = 3) of per cent stimulation/inhibition values of enzyme activity relative to untreated control

[a]Experimental per cent values (first row) over control

[b]Expected per cent values (second row) over control

A: *Antagonistic* insecticide interaction

B: *Additive* insecticide interaction

C: *Synergistic* insecticide interaction

Table 6.2 Interaction effects of insecticide combinations on invertase activity* in soil amended with N-P-K

		Buprofezin (µg g⁻¹ soil)				
		0	**2.5**	**5**	**7.5**	**10**
Acephate (µg g⁻¹ soil)	**0**	Control	0.71 ± 0.07 9^a	0.82 ± 0.13 26	0.72 ± 0.28 11	0.64 ± 0.01 −1
	2.5	65 ± 0.25 1	0.91 ± 0.3^C 40^a 10^b	1.0 ± 0.3^C 54 27	0.89 ± 0.1^C 37 12	0.53 ± 0.15^A −18 0.01
	5	0.9 ± 0.26 40	1.63 ± 0.36^C 151 45	1.06 ± 0.06^C 63 56	0.79 ± 0.01^A 21 47	0.52 ± 0.06^A −20 39
	7.5	0.69 ± 0.01 6	0.78 ± 0.01^C 20 14	0.54 ± 0.13^A −17 30	0.72 ± 0.02^A 11 16	0.56 ± 0.15^A −14 5
	10	0.54 ± 0.07 −17	0.69 ± 0.15^A 6 −6	0.57 ± 0.2^A −21 13	0.66 ± 0.05^A 1 −4	0.56 ± 0.06^A −14 −18

*mg glucose g⁻¹ 6 h⁻¹

Control value, 0.65 ± 0.15 mg glucose g⁻¹ 6 h⁻¹

All entries are means (n = 3) of per cent stimulation/inhibition values of enzyme activity relative to untreated control

[a]Experimental per cent values (first row) over control

[b]Expected per cent values (second row) over control

A: *Antagonistic* insecticide interaction

B: *Additive* insecticide interaction

C: *Synergistic* insecticide interaction

References

Antonions GF (2003) Impact of soil management and two botanical insecticides on urease and invertase activity. J Environ Sci Health B38:479–488

Begum SFM, Rajesh G (2015) Impact of microbial diversity and soil enzymatic activity in dimethoate amended soils series of Tamil Nadu. Int J Environ Sci Technol 4:1089–1097

Bielinska EJ, Pranagal J (2007) Enzymatic activity of soil contaminated with triazine herbicides. Pol J Environ Stud 16:295–300

Deborah BV, Mohiddin MJ, Madhuri RJ (2013) Interaction effects of selected pesticides on soil enzymes. Toxicol Int 20:195–200

Gianfreda L, Sannino F (1993) Influence of pesticides on the activity and kinetics of invertase, urease and acid phosphatase enzymes. Pestic Sci 39:237–244

Lodhi A, Malik NN, Mahmood T, Azam F (2000) Response of soil microflora, microbial biomass and some soil enzymes to Baythroid (an insecticide). Pak J Biol Sci 3:868–871

Mishra PC, Pradhan SC (1987) Seasonal variation in amylase, invertase, cellulase activity and carbon dioxide evolution in tropical protected grassland of Orissa, India, sprayed with carbaryl insecticide. Environ Pollut 43:291–300

Sannino F, Gianfreda L (2001) Pesticide influence on soil enzymatic activities. Chemosphere 45:417–425

Schäffer A (1993) Pesticide effects on enzyme activities in the soil ecosystem. In: Bollag JM, Stotzky G (eds) Soil biochemistry, vol 8. Marcel Dekker, New York, pp 273–340

Srinivasulu M, Rangaswamy V (2006) Activities of invertase and cellulase as influenced by the application of tridemorph and captan to ground nut (*Araachis hypogeae*) soil. Afr J Biotechnol 5:175–180

Tu CM (1982) Influence of pesticides on activities of amylase, invertase and level of adenosine triphosphate in organic soil. Chemosphere 2:909–914

Tu CM (1993) Effect of fungicides, captafol and chlorothalonil, on microbial and enzymatic activities in mineral soil. J Environ Sci Health B28:67–80

Voets JP, Meerschman P, Verstraete W (1974) Soil microbiological and biochemical effects of long-term atrazine applications. Soil Biol Biochem 6:149–152

Walia A, Mehta P, Guleria S, Chauhan A, Shirkot CK (2014) Impact of fungicide mancozeb at different application rates on soil microbial populations, soil biological processes, and enzyme activities in soil. Sci World J 2014:1–9

Wang C, Wang F, Zhang Q, Liang W (2016) Individual and combined effects of tebuconazole and carbendazim on soil microbial activity. Eur J Soil Biol 72:6–13

Wang C, Zhang Q, Wang F, Liang W (2017) Toxicological effects of dimethomorph on soil enzymatic activity and soil earthworm (*Eisenia fetida*). Chemosphere 169:316–323

Xiang HW, Li ZF, Xia TH (2009) Effect of omethoate on soil enzyme activities. J Scientia Agric Sinica 42:4282–4287

Xiong D, Gao Z, Fu B, Sun H, Tian S, Xiao Y, Qin Z (2013) Effect of pyrimorph on soil enzymatic activities and respiration. Eur J Soil Biol 56:44–48

Yan H, Wang D, Dong B, Tang F, Wang B, Fang H, Yu Y (2011) Dissipation of carbendazim and chloramphenicol alone and in combination and their effects on soil fungal:bacterial ratios and soil enzyme activities. Chemosphere 84:634–641

Yong-hong LI, Yu-bao GAO (2005) Effects of Monosulfuron on respiration, hydrogenase and invertase activity in soil. J Agro-environment Sci 24:1176–1181

Yu YL, Shan M, Fang H, Wang X, Chu XO (2006) Responses of soil microorganisms and enzymes to repeated applications of chlorothalonil. J Agric Food Chem 54:10070–10075

Chapter 7
Impact of Acephate and Buprofezin on Soil Proteases

Assay of Proteases in Soil

The activity of proteases in soil samples was determined by the method of Speir and Ross (1975). Soil samples (5 g) were incubated for 24 h at 30 °C with 10 mL of 0.1 M Tris (2-amino-2-(hydroxymethyl)propane-1,3-diol, pH 7.5) containing sodium caseinate (2% w/v). Four milliliters of aqueous solution (17.5%, w/v) of trichloro acetic acid was then added, and the mixture was centrifuged. A suitable aliquot of the supernatant was treated with 3 mL of 1.4 M Na_2CO_3 followed by the addition of 1 mL Folin-Ciocalteau reagent (33.3%, v/v). The blue color developed was read after 30 min at 700 nm in a spectrophotometer. Tyrosine was used as a standard. The activity of proteases was expressed as milligrams of tyrosine released per g of soil in 24 h (mg tyrosine g^{-1} soil 24 h^{-1}).

Nontarget Effects on Proteases

The data on protease activity in soil as influenced by application of acephate and buprofezin are shown in Figs. 7.1a, b, respectively. Single application of acephate, even at 7.5 µg g^{-1} concentration, stimulated (51%) protease activity in soil (Fig. 7.1a). However, after two or three applications of the insecticide, activity of proteases was stimulated up to 5.0 µg g^{-1} concentrations. Application of higher doses of acephate resulted in gradual decrease in the activity of proteases. On the other hand, the response of protease to buprofezin application was very similar to that of acephate (Fig. 7.1b). Concentrations up to 7.5 µg of buprofezin g^{-1} soil were either nontoxic or stimulatory to protease activity after a single application (Fig. 7.1b). Furthermore, highest activity of the enzyme was noticed in soil samples that received buprofezin

N.R. Maddela, K. Venkateswarlu, *Insecticides–Soil Microbiota Interactions*,
DOI 10.1007/978-3-319-66589-4_7

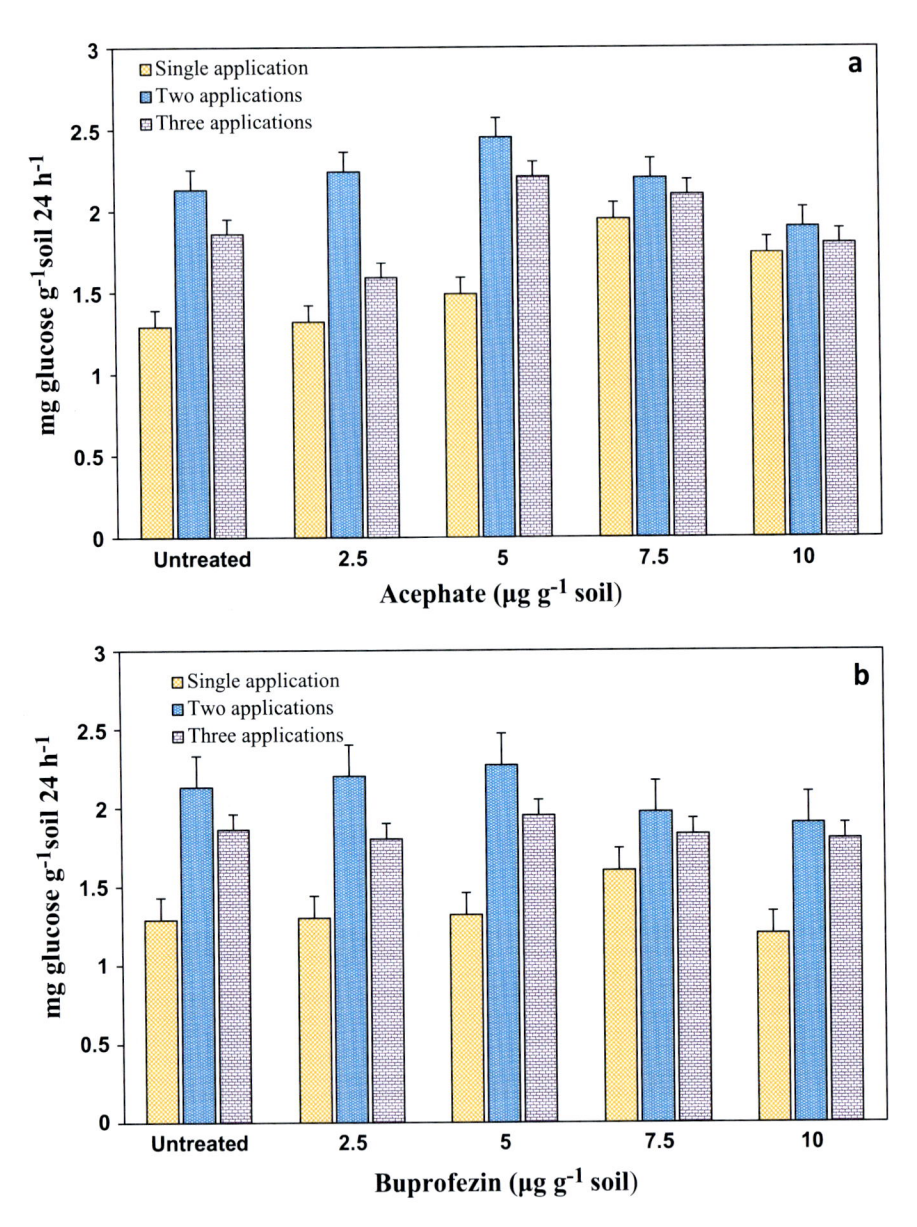

Fig. 7.1 Effect of single and repeated applications of (**a**) acephate and (**b**) buprofezin on activity of proteases in soil. Error bars represent standard deviations (n = 3)

at 5 µg g^{-1} soil. However, the enzyme activity declined rapidly at the highest concentration of buprofezin used (10 µg g^{-1}). Interestingly, the stimulatory effect of buprofezin continued even after two repeated applications. The activity of proteases decreased slightly after three applications of the insecticide to soil.

In agricultural soils, applications of various pesticides caused significant changes in the activities of soil proteases. According to Rangaswamy et al. (1994), monocrotophos, cypermethrin or fenvalerate even at 10 kg ha^{-1} were stimulatory to protease activity; however, the enzyme activity was negatively affected by the three insecticides at 12.5 kg ha^{-1}. Newell et al. (1981) studied the impact after 24–72-h exposure of an organophosphorous insecticide, fenthion (10^{1}–10^{3} ppb) on a fungal community, nitrogen-fixing microbes, and representative microfaunal and zooplankton invertebrates of a mangrove ecosystem. Acute lethal, growth-inhibiting, or process disrupting effects were not detected with exposures to less than 500 ppb fenthion. In a very recent investigation, Sanchez-Hernandez et al. (2017) reported that the activity of proteases was adversely affected by chlorpyrifos in 2 weeks. The activity decreased by 16 and 19% when soil was treated with 4.8 and 24 kg a.i. ha^{-1} of chlorpyrifos, respectively. Similarly, soil treated with metsulfuron-methyl (Ismail et al. 1998) exhibited short-lived inhibitory effect on protease activity in soil. Conversely, decreased protease activity was observed in soils treated with herbicides (Pahwa and Bajaj 1999), insecticides (Omar and Abd-Alla 2000) and chlorothalonil (Singh et al. 2002). Rasool and Reshi (2010) also observed that mancozeb applications had a stimulatory effect on protease activity in soil in comparison with control at lower concentrations. However, Endo et al. (1982) suggested that depression in protease activity in soil treated with 100 and 1000 µg g^{-1} soil of cartap hydrochloride under upland conditions.

Interestingly, Ismail et al. (1998) observed that protease activity in either Sungai Buluh or lating series soil treated with 5 µg g^{-1} soil of metsulfuron-methyl decreased during 7 days of incubation when compared with that of untreated control; however, a trend for recovery was observed with all the concentrations from day 14 onwards. Similar results were observed in more recently while working on the effect of a novel chiral insecticide, paichongding (IPP), on activity of proteases in two paddy soils (Chen et al. 2017). Though the enzyme response was differential in two soils in the early days of incubation, activities in both the soils were very similar in the later stages. Masko et al. (1991) observed no significant effect of phenmedipham even at 10 mg kg^{-1} on protease activity in soil.

On the other hand, detailed field experiments with herbicides also provide some useful information to understand the responses of proteases in soil. For instance, a long-term (17 months) field-level experiment was conducted by Oleszczuk et al. (2014) to understand the bevior of proteases in soils treated with two herbicides (2,4-D and Dicamba) and biochar individually and in combination. Selected herbicides showed different effects on soil proteases. Thus, 2,4-D at the applied dose (0.4 L ha^{-1}) decreased the enzyme activity by 49%; however, the negative effects of the herbicide on proteases were substantially alleviated with the application of 45 t ha^{-1} of biochar. This may be due to the sorption properties of biochar that decreased the bioavailability and toxicity of 2,4-D in soil. The stimulation in the enzyme activity was drastically decreased by higher doses of biochar, indicating that excess nutrient application does not help in enhancing soil fertility. Similar research focused on another herbicide (butachlor) versus proteases in paddy fields (Rasool et al. 2014). Under un-flooded conditions, activity of proteases decreased significantly after 14 days of incubation with all the applied doses of butachlor excepting a 10-fold concentration (15 kg a.i. ha^{-1}) wherein the activity was stimulated by 6.7%. Furthermore, irrespective of the concentration of butachlor, activity of proteases increased substantially after 21 days of application. In contrast, all the concentrations of butachlor stimulated protease activity under flooded soil conditions. Thus, the available literature and results of the present investigation clearly suggest that proteases respond negatively to pesticides at concentrations above the field application rates.

The data on behavior of proteases in soil samples that received NPK-fertilizer together with insecticide treatment are shown in Fig. 7.2. Acephate, up to 7.5 µg g^{-1}, when applied singly or twice, could stimulate (32%) the activity of proteases (Fig. 7.2a). However, upon three repeated applications at 10 µg g^{-1} soil, the enzyme activity was adversely affected (43% decline) in amended soil. Similarly, buprofezin application influenced soil proteases (Fig. 7.2b). Single application of buprofezin, up to 7.5 µg g^{-1}, was either nontoxic or stimulatory (17%) to protease in soil. The insecticide only at 10 µg g^{-1} soil inhibited (24%) the activity of protease. As with unamended soil, two applications of buprofezin also stimulated the protease activity in NPK-fertilizer amended soil. However, the enzyme activity markedly declined after three repeated applications. Renella et al. (2005) performed long-term field experiments in which soil, contaminated with Mn-Zn- or Cd-Ni-rich sludge, was incorporated at two different rates. Protease activity was generally more pronounced in all the sludge-amended soils than in control soils.

The two insecticides, in combination, yielded antagonistic, additive and synergistic interactions on protease activity in soil (Table 7.1). Acephate and buprofezin at concentrations of 5 µg g^{-1} soil exerted a synergistic increase (44% stimulation) in protease activity after 3 days incubation. Acephate treatment, at 5.0 µg g^{-1} soil, in combination with 2.5 µg g^{-1} of buprofezin exhibited additive response. However, higher concentrations from 5 µg g^{-1} soil of acephate or buprofezin were significantly toxic resulting in antagonism towards the enzyme activity. On the other hand, fertilizer amendments to the soil samples together with the insecticide combinations at graded levels strongly inhibited the enzyme activity resulting in antagonism (Table 7.2). Surprisingly, 5 µg g^{-1} soil of acephate when combined with buprofezin at concentrations ranging from 5–10 µg g^{-1} soil interacted antagonistically with protease in NPK-amended soil. Thus, in contrast to the response of invertase to the insecticides, proteases are significantly affected by the insecticides in soil amended with N-P-K fertilizers.

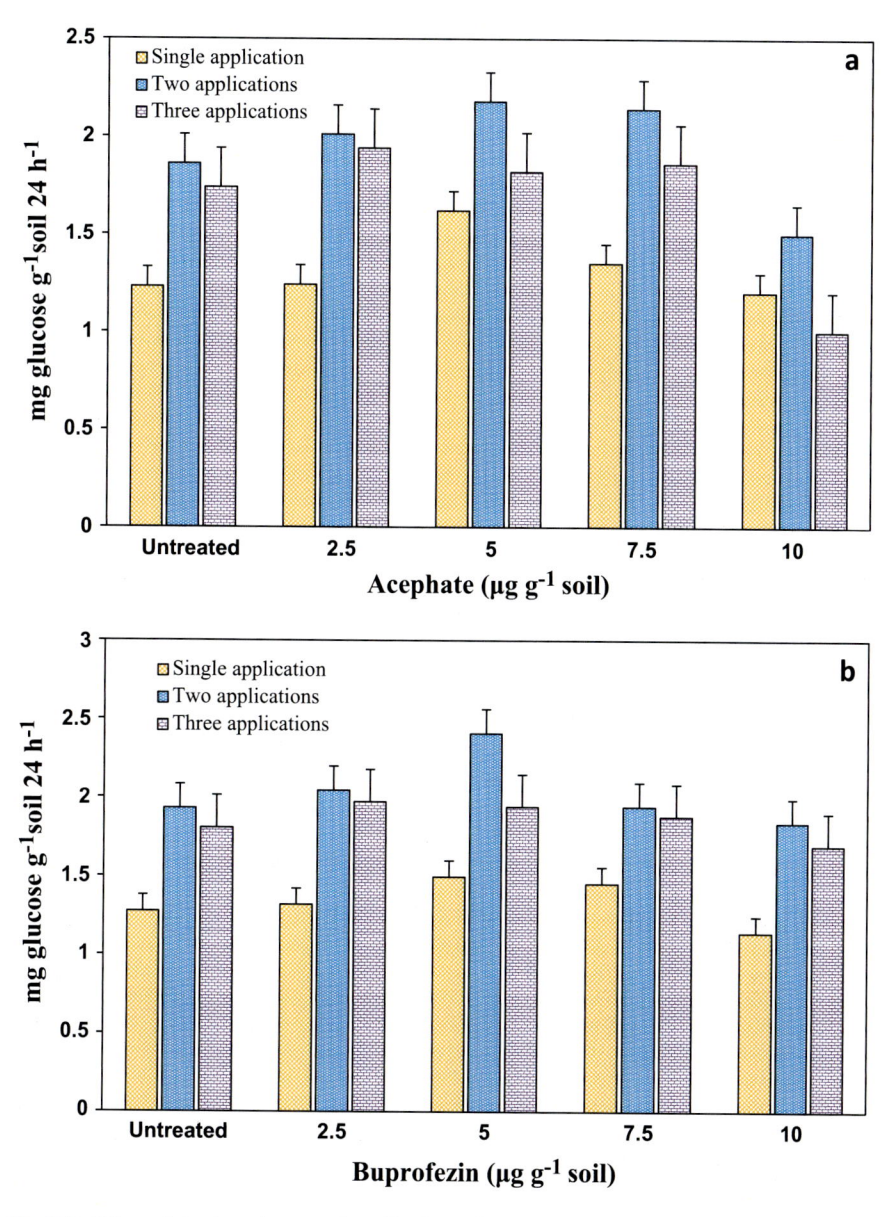

Fig. 7.2 Effect of single and repeated applications of (**a**) acephate and (**b**) buprofezin on activity of proteases in soil amended with N-P-K. Error bars represent standard deviations (n = 3)

Table 7.1 Interaction effects of insecticide combinations on activity of proteases* in soil

Acephate (µg g^{-1} soil)	Buprofezin (µg g^{-1} soil)				
	0	**2.5**	**5**	**7.5**	**10**
0	Control	1.30 ± 0.05 1[a]	1.32 ± 0.1 2	1.60 ± 0.15 24	1.31 ± 0.03 1
2.5	1.32 ± 0.06 2	1.37 ± 0.11[C] 6[a] 3[b]	1.35 ± 0.06[C] 5 4	1.11 ± 0.03[A] −14 29	1.07 ± 0.04[A] −17 7
5	1.49 ± 0.16 15	1.54 ± 0.07[B] 19 19	1.86 ± 0.05[C] 44 17	1.48 ± 0.15[A] 15 36	1.36 ± 0.16[A] 5 16
7.5	1.95 ± 0.16 52	2.01 ± 0.05[C] 56 53	1.48 ± 0.2[A] 15 53	1.44 ± 0.01[A] 12 63	1.55 ± 0.14[A] 20 52
10	1.74 ± 0.18 34	1.54 ± 0.12[A] 19.38 37	1.55 ± 0.1[A] 20.15 35	1.59 ± 0.14[A] 23.25 50	1.44 ± 0.23[A] 11.63 35

* mg tyrosine g^{-1} 24 h^{-1}

Control value, 1.29 ± 0.14 mg tyrosine g^{-1} 24 h^{-1}

All entries are means (n = 3) of per cent stimulation/inhibition values of enzyme activity relative to untreated control

[a]Experimental per cent values (first row) over control

[b]Expected per cent values (second row) over control

A: *Antagonistic* insecticide interaction

B: *Additive* insecticide interaction

C: *Synergistic* insecticide interaction

Table 7.2 Interaction effects of insecticide combinations on activity of proteases* in soil amended with N-P-K

		Buprofezin (µg g⁻¹ soil)				
		0	**2.5**	**5**	**7.5**	**10**
Acephate (µg g⁻¹ soil)	**0**	Control	1.27 ± 0.03 3^a	1.44 ± 0.04 17	1.40 ± 0.15 14	1.37 ± 0.18 11
	2.5	1.24 ± 0.06 1	1.34 ± 0.04^C 9^a 4^b	1.8 ± 0.03^C 46 18	1.45 ± 0.03^C 18 15	0.97 ± 0.04^A -21 12
	5	1.62 ± 0.21 32	1.7 ± 0.09^C 38 34	0.57 ± 0.03^A -54 44	0.42 ± 0.1^A -66 41	0.93 ± 0.12^A -24 39
	7.5	1.35 ± 0.14 10	1.42 ± 0.01^C 15 13	0.34 ± 0.04^A -72 25	0.43 ± 0.03^A -65 23	0.46 ± 0.01^A -63 20
	10	1.25 ± 0.08 2	1.13 ± 0.07^A -8 5	0.52 ± 0.02^A -58 11	0.52 ± 0.02^A -58 15	0.56 ± 0.04^A -54 13

* mg tyrosine g⁻¹ 24 h⁻¹

Control value, 1.23 ± 0.27 mg tyrosine g⁻¹ 24 h⁻¹

All entries are means (n = 3) of per cent stimulation/inhibition values of enzyme activity relative to untreated control

[a]Experimental per cent values (first row) over control

[b]Expected per cent values (second row) over control

A: *Antagonistic* insecticide interaction

B: *Additive* insecticide interaction

C: *Synergistic* insecticide interaction

References

Chen J, Zhou S, Rong Y, Zhu X, Zhao X, Cai Z (2017) Pyrosequencing reveals bacterial communities and enzyme activities differences after application of novel chiral insecticide Paichongding in aerobic soils. Appl Soil Ecol 112:18–27

Endo T, Kuska T, Tan N, Sakai M (1982) Effects of the insecticide Cartap hydrochloride on soil enzyme activities, respiration and nitrification. J Pestic Sci 7:101–110

Ismail BS, Yapp KF, Omar O (1998) Effects of metsulfuron-methyl on amylase, urease and protease activities in two soils. Aust J Soil Res 36:449–456

Masko AA, Lovehii NF, Pototskaya LA (1991) Stability of immobilized soil enzymes and their role in the degradation of herbicides. Akad Village, Sciences Byelorussian Soviet Socialist Republic, Ser Biol Sci (Vestsi Akad. Navuk BSSR, Ser. Biyal Navuk) 5:47–51

Newell SY, Cooksey KE, Fell JW, Master JM, Miller C, Walter MA (1981) Acute impact of an organophosphorus insecticide on microbes and small invertebrates of a mangrove estuary. Arch Environ Contam Toxicol 10:427–435

Oleszczuk P, Jośko I, Futa B, Pasieczna-Patkowska S, Pałys E, Kraska P (2014) Effect of pesticides on microorganisms, enzymatic activity and plant in biochar-amended soil. Geoderma 214–215:10–18

Omar SA, Abd-Alla MA (2000) Microbial populations and enzyme activities in soil treated with pesticides in Egypt. Water Air Soil Pollut 127:49–63

Pahwa SH, Bajaj K (1999) Effect of pre-emergence herbicides on the activity of α-amylase and protease enzyme during germination in pigeon pea and carpet weed. Indian J Weed Sci 31:148–150

Rangaswamy V, Reddy BR, Venkateswarlu K (1994) Activities of dehydrogenase and protease in soil as influenced by monocrotophos, quinalphos, cypermethrin and fenvalerate. Agric Ecosyst Environ 47:319–326

Rasool N, Reshi ZA (2010) Effect of the fungicide mancozeb at different application rates on enzyme activities in a silt loam of the Kashmir Himalaya, India. Trop Ecol 51:199–205

Rasool N, Reshi ZA, Shah MA (2014) Effect of butachlor (G) on soil enzyme activity. Eur J Soil Biol 61:94–100

Renella G, Mench M, Gelsomino A, Landi L, Nannipieri P (2005) Functional activity and microbial community structure in soils amended with bimetallic sludges. Soil Biol Biochem 37:1498–1506

Sanchez-Hernandez JC, Sandoval M, Pierart A (2017) Short-term response of soil enzyme activities in a chlorpyrifos-treated mesocosm: use of enzyme-based indexes. Ecol Indic 73:525–535

Singh BK, Allan W, Denus JW (2002) Degradation of chlorpyrifos, fenamiphos and chlorothalonil alone and in combination and their effects on soil microbial activity. Environ Toxicol Chem 21:2600–2605

Speir TW, Ross DJ (1975) Effects of storage on the activities of protease, urease, phosphatase and sulphatase in three soils under pasture. NZ J Sci 18:231–237

Chapter 8
Impact of Acephate and Buprofezin on Soil Urease

Assay of Urease in Soil

Urease activity in untreated and insecticide- and/or fertilizer-treated soil samples (5 g) was determined by incubating for 30 min at 30 °C with 4 mL of 0.2 M sodium phosphate buffer (pH 7.0) and 1 mL of 1 M urea (prepared in sodium phosphate buffer). Ten milliliters of aqueous solution of potassium chloride (2 M) was added, and the mixture was kept at 4 °C for 10 min and centrifuged (Zantua and Bremner 1975). A suitable aliquot of the supernatant was treated with 0.5 mL of Nessler's reagent followed by the addition of 3.5 mL of distilled water. The red color developed was read at 495 nm in a digital spectrophotometer. By using ammonium sulphate as a standard, urease activity was expressed in terms of milligrams of ammonical nitrogen released per g of soil in 30 min (mg NH_4^+-N released g^{-1} soil 30 m^{-1}).

Nontarget Effects on Urease

Soil urease, a vital enzyme in nitrogen turnover, has been assessed for the nontarget effects of single or repeated applications of acephate or buprofezin by determining the ammonical nitrogen released from urea, and the results are presented in Fig. 8.1. Though the activity was stimulated (10–40%) at concentrations of 2.5 and 5 $\mu g\ g^{-1}$, the enzyme was affected adversely (20–30% inhibition) at 7.5 and 10 μg acephate g^{-1} soil (Fig. 8.1a). Again, activity was stimulated after two repeated applications of acephate, but was inhibited after three applications to soil. On the contrary, buprofezin was much lesser toxic to urease than acephate (Fig. 8.1b). Buprofezin, at concentrations ranging from 2.5 to 7.5 $\mu g\ g^{-1}$ soil was either stimulatory or nontoxic to urease. Surprisingly, even 7.5 μg buprofezin g^{-1} soil did not affect urease in soil.

© Springer International Publishing AG 2018
N.R. Maddela, K. Venkateswarlu, *Insecticides–Soil Microbiota Interactions*,
DOI 10.1007/978-3-319-66589-4_8

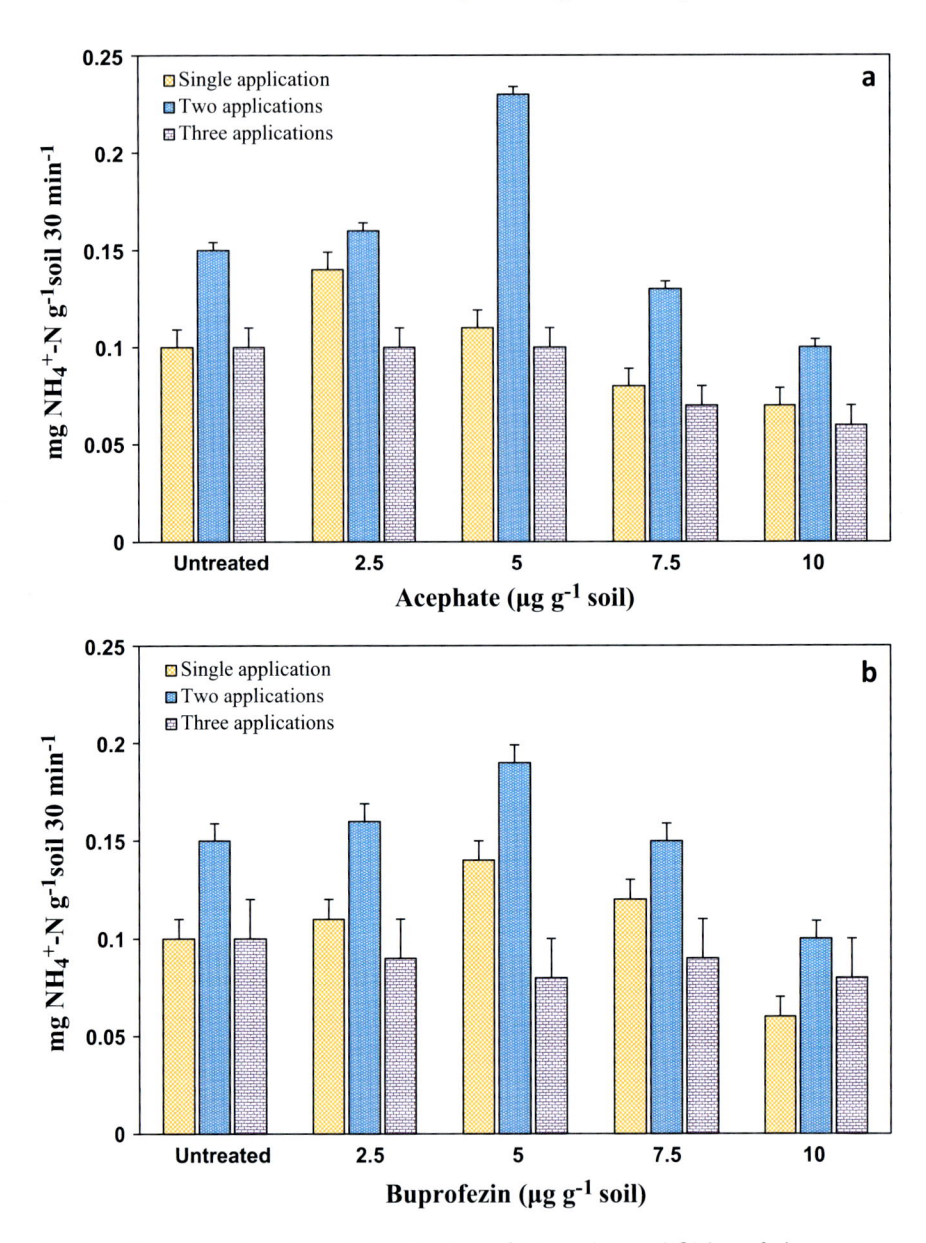

Fig. 8.1 Effect of single and repeated applications of (**a**) acephate and (**b**) buprofezin on urease activity in soil. Error bars represent standard deviations (n = 3)

Maximum enzyme activity (40% increases) was recorded in soil samples that received 5 µg g^{-1} of buprofezin. The activity was stimulated (7–27%) even after two applications of buprofezin at lower rates to soil. However, urease activity decreased by 30–50% after three repeated applications of the insecticide.

It is well documented that urease activity in soils is influenced by many factors which include cropping history, organic amendments, heavy metals, and environmental factors such as temperature (Tabatabai 1977). Nevertheless, the agricultural chemicals especially pesticides are perhaps the largest groups of poisonous substances being disseminated throughout the environment. Available evidences also suggest that ureolytic microorganisms isolated from soil were inhibited by OP pesticides to a greater or lesser extent, but the development of tolerance was common (Lethbridge and Burns 1976). Lower concentration (50 µg g^{-1}) of three organophosphorus insecticides, viz., malathion, accothion and thimet was also inhibitory for the enzyme in sand clay loam soil (Lethbridge and Burns 1976). Application of 1000 µg g^{-1} insecticide to soil resulted in 40% inhibition of urea hydrolysis after 60 days for accothion, and 50% for malathion and thimet. Similar inhibitory effects were recorded in a silt loam soil wherein 200 µg g^{-1} applications exhibited inhibition ranging from 14% (accothion) to 23% (thimet) after 10 days. Metsulfuron-methyl at 5 µg g^{-1} caused a reduction in urease activity up to 28 days in in loamy sand and clay loam soil (Ismail et al. 1998). Kalam et al. (2004) suggested that soil urease activity was affected markedly in presence of profenofos and the inhibition was 62% at 1000 mg kg^{-1} level after 80 days. Similarly, soils treated with diazinon (Ingram et al. 2005), acetamiprid (Singh and Kumar 2008), and omethoate (Xiang et al. 2009) significantly inhibited soil urease activity.

Application of chlorothalonil (Yu et al. 2006), amitraz, tebupirimphos and aztec (Tu 1995) and dimethomorph (Wu et al. 2010) to soils showed inhibitory effects for only short period of time. On the contrary, in soils treated with baythroid (Lodhi et al. 2000), glyphosate and paraquat (Sannino and Gianfreda 2001), chlorimuron-ethyl and furandan (Yang et al. 2006) the rates of urease activities were significantly higher. Even 100 mg fenamiphos kg^{-1} soil was not toxic to urease (Caceres et al. 2009). Tu (1995) suggested that amitraz, tebupirimphos and aztec inhibited urease activity in soil after a week. Metsulfuron-methyl, at 5 µg g^{-1}, caused a reduction in urease activity at 28 days of incubation (Ismail et al. 1998). Activity of urease was adversely affected by dimethoate during the entire incubation period (Begum and Rajesh 2015). The enzyme activity increased gradually from day 0 to 60 and then dropped until day 120 both in controls and dimethoate-treated samples, indicating that urease is highly sensitive to dimethoate. Small increases were measured for urease in soils treated with glyphosate and paraquat (Sannino and Gianfreda 2001). Similarly, Yang et al. (2006) reported that chlorimuron-ethyl and Furadan could activate urease activity of test soils, with the largest increments of 46.95% and 39.36% by chlorimuron-ethyl, and 21.08% and 12.70% by Furadan in meadow brown soil and black soil, respectively. Lodhi et al. (2000) observed a maximum increase of 40.9% in urease activity at 0.8 µg g^{-1} baythroid in soil, while higher levels were inhibitory to the enzyme as the decline was substantial to 9.1%. Similary, a fungicide, dimethomorph stimulated urease activity marginally at lower concentrations

(1.0 and 10.0 mg kg^{-1}), but inhibited the enzyme activity (5–29%) at higher dose (100 mg kg^{-1}) (Wang et al. 2017). Acetamiprid (neonicotinoid insecticide) strongly inhibited (35% reduction) urease in soil (Singh and Kumar 2008).

Urease enzyme responded differently in soil samples treated with another neonicotinoid insecticide, imidacloprid, both at field (1 mg kg^{-1} soil) and higher (10 mg kg^{-1} soil) rates (Cycoń and Piotrowska-Seget 2015). After 24 h of application, the enzyme activity declined by 10 and 30% in soils that received 1.0 and 10 mg imidacloprid, respectively. In contrast, 2–8 weeks after the treatment, the increase in enzyme activity was 7–13% and 18–35% in soil samples that received 1 and 10 mg of the insecticide, respectively. These results clearly suggest that urease producers or urease enzyme are stable against long-term exposure to insecticides such as imidacloprid. Yu et al. (2006) reported that the urease activity decreased significantly after the first and second additions of chlorothalonil to soil, but the negative effects became transient and weaker following the third and fourth applications. The impact of a chiral insecticide, paichongding (IPP), on urease was different in desalting muddy polder and yellow paddy soil (Chen et al. 2017). The enzyme activity increased between 20 and 45 days after treatment in muddy polder, and was inhibited later by the insecticide. Voets et al. (1974) reported that long-term atrazine applications significantly reduced the activity of urease in soil. Jastrzebska and Kucharski (2007) demonstrated that increasing doses of two fungicides significantly inhibited the activity of urease, which was particularly noticeable in soil cropped with spring barley.

On the other hand, there are some compounds which either stimulated or unaffected the activity of urease in soil. For example, in a short-term field-scale study (14-day) using a natural insecticide derived from neem plant, azadirachtin, an increase in soil urease activity was observed at 0.6 L azadirachtin ha^{-1} soil (Kizilkaya et al. 2015). Chlorpyrifos did not exhibit any adverse effects on urease activity in soil at concentrations of 4.8 and 24 kg a.i. ha^{-1} (Sanchez-Hernandez et al. 2017). Cypermethrin showed marginal negative effect (3%) on soil urease activity over control (Filimon et al. 2015).

The response of urease was quite interesting in soil samples that received both the insecticides and N-P-K amendments. Urease activity was significantly enhanced (20–113%) with a single application of acephate at all four concentrations tested (2.5–10 μg g^{-1}) in NPK-amended soil (Fig. 8.2a). Most pronounced activity was noticed in soil that was treated with the insecticide at 5 μg g^{-1} soil. Furthermore, repeated applications of acephate to soil amended with NPK-fertilizer caused a significant reduction of 70–81% in urease activity. In fact, urease activity was decreased by more than 50% after two applications over the activity recorded after a single application. Again, the lowest activity was noticed after three repeated applications. Soil urease responded to buprofezin very similarly to that of acephate in NPK-amended soil (Fig. 8.2b).

Ingram et al. (2005) applied diazinon and imidacloprid to lawns for insect control simultaneously with nitrogenous fertilizer such as urea and observed that diazinon briefly, but significantly, reduced urease activity in blue grass sod. Co-application of imidacloprid and urea appeared to increase urease activity in soil and sod. The response of urease against 2,4-D alone and in combination with biochar (carbon-rich substance) was significantly different (Oleszczuk et al. 2014). 2,4-D alone did not

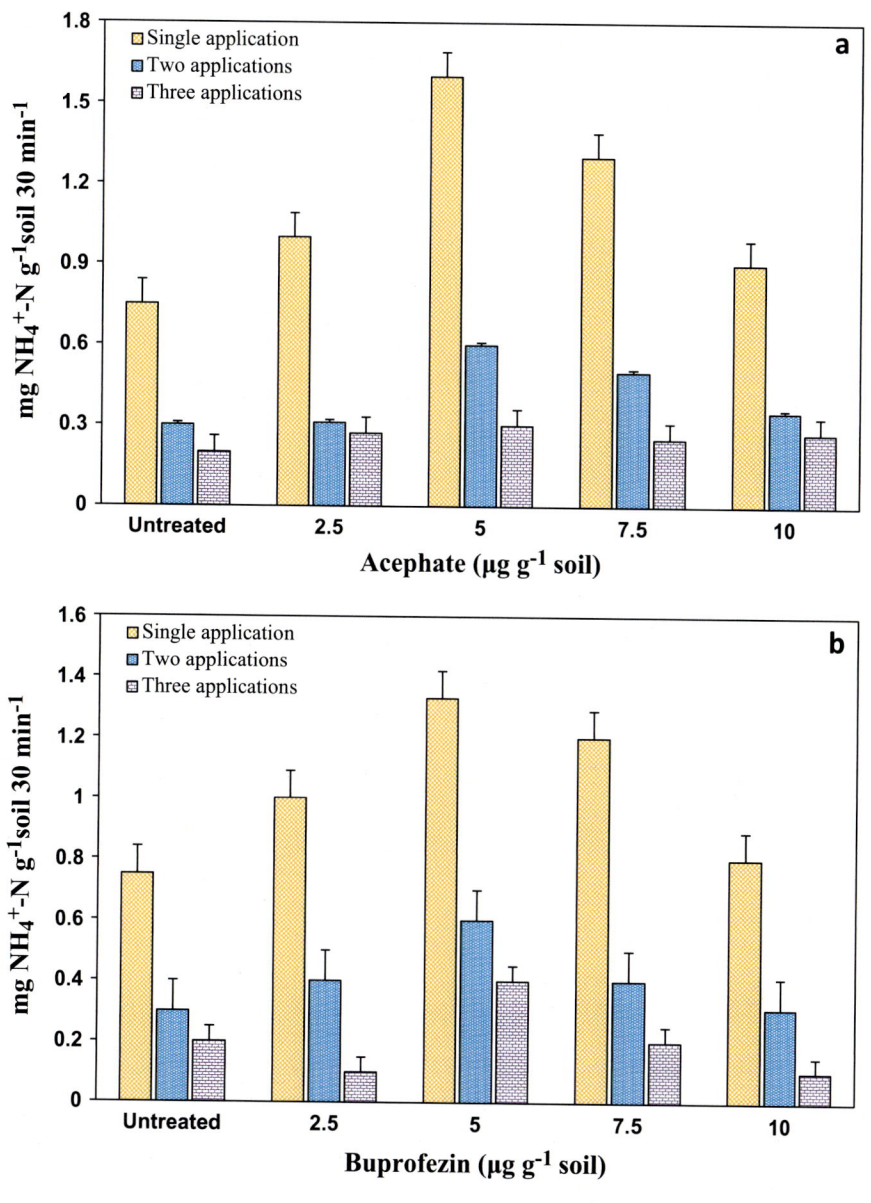

Fig. 8.2 Effect of single and repeated applications of (**a**) acephate and (**b**) buprofezin on urease activity in soil amended with N-P-K. Error bars represent standard deviations (n = 3)

show any impact on soil urease activity. But, application of biochar (30 t ha^{-1}) plus 2,4-D caused significant adverse effect on enzyme activity over control soil (with no 2,4-D and with a biochar). Furthermore, the activity of urease was well protected against dicamba by biochar (45 t ha^{-1}), indicating that the affinity of a fertilizer to pesticide varies with their combinations. Also, Bielinska and Pranagal (2007)

Table 8.1 Interaction effects of insecticide combinations on urease activity* in soil

		Buprofezin (μg g^{-1} soil)				
		0	**2.5**	**5**	**7.5**	**10**
Acephate (μg g^{-1} soil)	**0**	Control	0.11 ± 0.01 10[a]	0.14 ± 0.01 40	0.12 ± 0.01 20	0.11 ± 0.04 10
	2.5	0.14 ± 0.01 40	0.15 ± 0.01[B] 50[a] 46[b]	0.18 ± 0.01[C] 80 64	0.12 ± 0.01[A] 20 52	0.09 ± 0.01[A] −10 46
	5	0.11 ± 0.01 10	0.15 ± 0.02[C] 50 19	0.19 ± 0.01[C] 90 46	0.14 ± 0.01[C] 40 28	0.09 ± 0.01[A] −10 19
	7.5	0.10 ± 0.01 1	0.14 ± 0.02[C] 40 11	0.12 ± 0.02[A] 20 40	0.10 ± 0.01[A] 1 21	0.10 ± 0.01[A] 1 11
	10	0.10 ± 0.02 1	0.09 ± 0.01[A] −10 11	0.12 ± 0.04[A] 20 40	0.10 ± 0.09[A] 1 21	0.09 ± 0.03[A] −10 11

*mg NH$_4^+$-N g^{-1} 30 min^{-1}

Control value, 0.10 ± 0.03 mg NH$_4^+$-N g^{-1} 30 min^{-1}

All entries are means (n = 3) of per cent stimulation/inhibition values of enzyme activity relative to untreated control

[a]Experimental per cent values (first row) over control

[b]Expected per cent values (second row) over control

A: *Antagonistic* insecticide interaction

B: *Additive* insecticide interaction

C: *Synergistic* insecticide interaction

opined that application of high level of mineral fertilization simultaneously with chemical weed control was particularly detrimental to biological activity of soil. Renella et al. (2005) studied the effects of Mn-Zn- or Cd-Ni-rich sludge on soil enzyme activities by performing long-term field experiments, and found that urease activity was little affected by sludge amendments. However, urease enzyme was found to be highly sensitive to Ag and Hg in soil (Chaperon and Sauve 2007). Studies by many authors (Frankenberger and Johanson 1982; Acosta-Martinez and Tabatabai 2000; Aon and Colaneri 2001) indicated that an increase in the concentration of hydrogen ions in soil has a negative effect on its enzyme activity.

The impact of the two selected insecticides in combination on soil urease activity was also assessed in soil treated without and with NPK-fertilizer amendments. Acephate and buprofezin at equal concentrations of 2.5 or 5 μg g^{-1}, in combination, caused either an additive or synergistic effect on urease activity (Table 8.1). But, combination of the

Table 8.2 Interaction effects of insecticide combinations on urease activity* in soil amended with N-P-K

		Buprofezin ($\mu g\ g^{-1}$ soil)				
		0	**2.5**	**5**	**7.5**	**10**
Acephate ($\mu g\ g^{-1}$ soil)	**0**	Control	1.68 ± 0.13 17^a	1.78 ± 0.10 24	1.71 ± 0.09 19	1.71 ± 0.2 19
	2.5	1.78 ± 0.06 24	1.95 ± 0.06^B 35^a 37^b	2.21 ± 0.19^C 53 42	2.21 ± 0.02^C 53 38	1.93 ± 0.14^A 34 38
	5	2.16 ± 0.1 50	2.5 ± 0.07^C 75 58	1.99 ± 0.21^A 38 62	2.01 ± 0.11^A 39 59	1.27 ± 0.32^A -11 59
	7.5	1.80 ± 0.28 25	2.34 ± 0.21^C 62 38	1.78 ± 0.06^A 24 43	1.71 ± 0.28^A 19 39	1.88 ± 0.16^A 30 39
	10	1.80 ± 0.15 25	1.95 ± 0.09^A 35 38	1.56 ± 0.08^A 8 50	1.96 ± 0.16^A 36 39	1.96 ± 0.04^A 36 39

*mg NH_4^+-N g^{-1} 30 min^{-1}

Control value, 1.44 ± 0.13 mg NH_4^+-N g^{-1} 30 min^{-1}

All entries are means (n = 3) of per cent stimulation/inhibition values of enzyme activity relative to untreated control

[a]Experimental per cent values (first row) over control

[b]Expected per cent values (second row) over control

A: *Antagonistic* insecticide interaction

B: *Additive* insecticide interaction

C: *Synergistic* insecticide interaction

insecticides at the higher level of either 7.5 or 10 $\mu g\ g^{-1}$ considerably reduced the urease activity resulting in antagonism. However, the activity of urease was adversely affected by the combination of the two insecticides in NPK-amended soil (Table 8.2).

Insecticide combinations even at 5 $\mu g\ g^{-1}$ soil also recorded antagonistic effect to soil urease. But, combination of the two insecticides at 2.5 $\mu g\ g^{-1}$ soil caused an additive response to soil urease activity. In contrast to the present findings, Rahman et al. (2003) reported that urease activity was higher in fertilizer amended soils than in non-amended soils. Similarly, other studies reported that urease activity in soil was not much affected by pesticides (Davies and Greaves 1981; Lethbridge et al. 1981). It is apparent from the above observations that nutrient amendments and insecticides treatment adversely affect soil urease activity. In particular, repeated applications of either acephate or buprofezin greatly inhibit the activity of urease in NPK-amended soils.

References

Acosta-Martinez V, Tabatabai MA (2000) Enzyme activities in a limed agricultural soil. Biol Fertil Soils 3:85–91

Aon MA, Colaneri AC (2001) Temporal and spatial evolution of enzymatic activities and physico-chemical properties in an agricultural soil. Appl Soil Ecol 18:255–270

Begum SFM, Rajesh G (2015) Impact of microbial diversity and soil enzymatic activity in dimethoate amended soils series of Tamil Nadu. Int J Environ Sci Technol 4:1089–1097

Bielinska EJ, Pranagal J (2007) Enzymatic activity of soil contaminated with triazine herbicides. Pol J Environ Stud 16:295–300

Caceres T, He W, Megharaj M, Naidu R (2009) Effect of insecticide fenamiphos on soil microbial activities in Australian and Ecuadorean soils. J Environ Sci Health B44:13–17

Chaperon S, Sauve S (2007) Toxicity interaction of metals (Ag, Cu, Hg, Zn) to urease and dehydrogenase activities in soils. Soil Biol Biochem 39:2329–2338

Chen J, Zhou S, Rong Y, Zhu X, Zhao X, Cai Z (2017) Pyrosequencing reveals bacterial communities and enzyme activities differences after application of novel chiral insecticide Paichongding in aerobic soils. Appl Soil Ecol 112:18–27

Cycoń M, Piotrowska-Seget Z (2015) Biochemical and microbial soil functioning after application of the insecticide imidacloprid. J Environ Sci 27:147–158

Davies HA, Greaves MP (1981) Effects of some pesticides on soil enzyme activities. Weed Res 21:205–209

Filimon MN, Voia SO, Popescu R, Dumitrescu G, Ciochina LP, Mituletu M, Vlad DC (2015) The effect of some insecticides on soil microorganisms based on enzymatic and bacteriological analyses. Rom Biotech Lett 20:10439–10447

Frankenberger WT, Johanson JB (1982) Effect of pH on enzyme stability in soils. Rom Biotech Lett 14:433–437

Ingram CW, Coyne MS, Williams DW (2005) Effects of commercial diazinon and imidacloprid on microbial urease activity in soil and sod. J Environ Qual 34:1573–1580

Ismail BS, Yapp KF, Omar O (1998) Effects of metsulfuron-methyl on amylase, urease and protease activities in two soils. Aust J Soil Res 36:449–456

Jastrzebska E, Kucharski J (2007) Dehydrogenases, urease and phosphatases activities of soil contaminated with fungicides. Plant Soil Environ 53:51–57

Kalam A, Tah J, Mukherjee AK (2004) Pesticide effects on microbial population and soil enzyme activities during vermicomposting of agricultural waste. J Environ Biol 25:201–208

Kizilkaya R, Samofalova I, Mudrykh N, Mikailsoy F, Akça I, Sushkova S, Minkina T (2015) Assessing the impact of azadirachtin application to soil on urease activity and its kinetic parameters. Turk J Agric For 39:976–983

Lethbridge G, Bull AT, Burns RG (1981) Effects of pesticides on 1,3-β-glucanase and urease activities in soils in the presence and absence of fertilizers, lime and organic materials. Pestic Sci 12:147–155

Lethbridge G, Burns RG (1976) Inhibition of soil urease by organophosphorus insecticides. Soil Biol Biochem 8:99–102

Lodhi A, Malik NN, Mahmood T, Azam F (2000) Response of soil microflora, microbial biomass and some soil enzymes to Baythroid (an insecticide). Pak J Biol Sci 3:868–871

Oleszczuk P, Jośko I, Futa B, Pasieczna-Patkowska S, Pałys E, Kraska P (2014) Effect of pesticides on microorganisms, enzymatic activity and plant in biochar-amended soil. Geoderma 214–215:10–18

Rahman MM, Kim T, Rhee IK, Kim JE (2003) Effect of the fungicide chlorothalonil on microbial activity and nitrogen dynamics in soil scosystem. Agric Chem Biotechnol 46:169–173

Renella G, Mench M, Gelsomino A, Landi L, Nannipieri P (2005) Functional activity and microbial community structure in soils amended with bimetallic sludges. Soil Biol Biochem 37:1498–1506

Sanchez-Hernandez JC, Sandoval M, Pierart A (2017) Short-term response of soil enzyme activities in a chlorpyrifos-treated mesocosm: use of enzyme-based indexes. Ecol Indic 73:525–535

Sannino F, Gianfreda L (2001) Pesticide influence on soil enzymatic activities. Chemosphere 45:417–425

Singh DK, Kumar S (2008) Nitrate reductase, arginine deaminase, urease and dehydrogenase activities in natural soil (ridges with forest) and in cotton soil after acetamiprid treatments. Chemosphere 71:412–418

Tabatabai MA (1977) Effects of trace elements on urease activity in soils. Soil Biol Biochem 9:9–13

Tu CM (1995) Effect of five insecticides on microbial and enzymatic activities in sandy soil. J Environ Sci Health B30:289–306

Voets JP, Meerschman P, Verstraete W (1974) Soil microbiological and biochemical effects of long-term atrazine applications. Soil Biol Biochem 6:149–152

Wang C, Zhang Q, Wang F, Liang W (2017) Toxicological effects of dimethomorph on soil enzymatic activity and soil earthworm (*Eisenia fetida*). Chemosphere 169:316–323

Wu Y, Kong F, Wu D, Yan Z, Chen Z, Deng T (2010) Effects of Dimethomorph on some enzymatic activities in soil. Bioinformatics and Biomedical Engineering (iCBBE), 14th International conference on 18–20 June. Chengdu, pp 1–3

Xiang HW, Li ZF, Xia TH (2009) Effect of omethoate on soil enzyme activities. J Scientia Agric Sinica 42:4282–4287

Yang C, Sun T, He W, Chen S (2006) Effects of pesticides on soil urease activity. J Appl Ecol 17:1354–1356

Yu YL, Shan M, Fang H, Wang X, Chu XO (2006) Responses of soil microorganisms and enzymes to repeated applications of chlorothalonil. J Agric Food Chem 54:10070–10075

Zantua MI, Bremner JM (1975) Comparison of methods of assaying urease activity in soils. Soil Biol Biochem 7:291–295

Chapter 9
Impact of Acephate and Buprofezin on Soil Phosphatases

Assay of Acid and Alkaline Phosphatases in Soil

Acid phosphatase activity in soil samples was determined following the method of Tabatabai and Bremner (1969). Soil samples (5 g) were incubated at 37 °C for 30 min with 15 mL of enzyme buffer (100 mM sodium acetate of pH 5.5 containing 10 mM $MgCl_2$) and 5 mL of 0.03 M p-nitrophenyl phosphate (PNPP). The mixture was kept on ice for 20 min and centrifuged. A suitable aliquot of the supernatant was treated with 3 mL of enzyme buffer, and was kept on ice for 20 min. Then, 1.0 mL of aqueous solution of 5 mM $CaCl_2$ and 4 mL of 0.5 M NaOH were added to the above mixture. The yellow color developed was read at 405 nm in a digital spectrophotometer. p-Nitrophenol (PNP) was used as a standard. Acid phosphatase activity was expressed as milligrams of PNP released from PNPP per g of soil in 30 min (mg PNP g^{-1} soil 30 min^{-1}). The activity of alkaline phosphatase in soils was determined following the method of Tabatabai and Bremner (1969) modified by Eivazi and Tabatabai (1977). The procedure was very similar to that adapted for assaying acid phosphatase excepting the inclusion of enzyme buffer (100 mM Tris-HCl of pH 8.6 containing 10 mM $MgCl_2$). The activity of alkaline phosphatases was also expressed as mg PNP g^{-1} soil 30 min^{-1}.

Nontarget Effects on Acid Phosphatase

The nontarget effects of acephate and buprofezin on acid phosphatase activity in soils were determined by the release of PNP from PNPP, and the results are presented in Fig. 9.1. It is evident from the data that after a single application, acephate at field rate was nontoxic or stimulatory to acid phosphatase activity in soil (Fig. 9.1a). The accumulation of PNP was more striking (43% stimulation) at 5 μg g^{-1} soil of acephate. The highest level of acephate (10 μg g^{-1} soil) was inhibitory (29%) to the

N.R. Maddela, K. Venkateswarlu, *Insecticides–Soil Microbiota Interactions*,
DOI 10.1007/978-3-319-66589-4_9

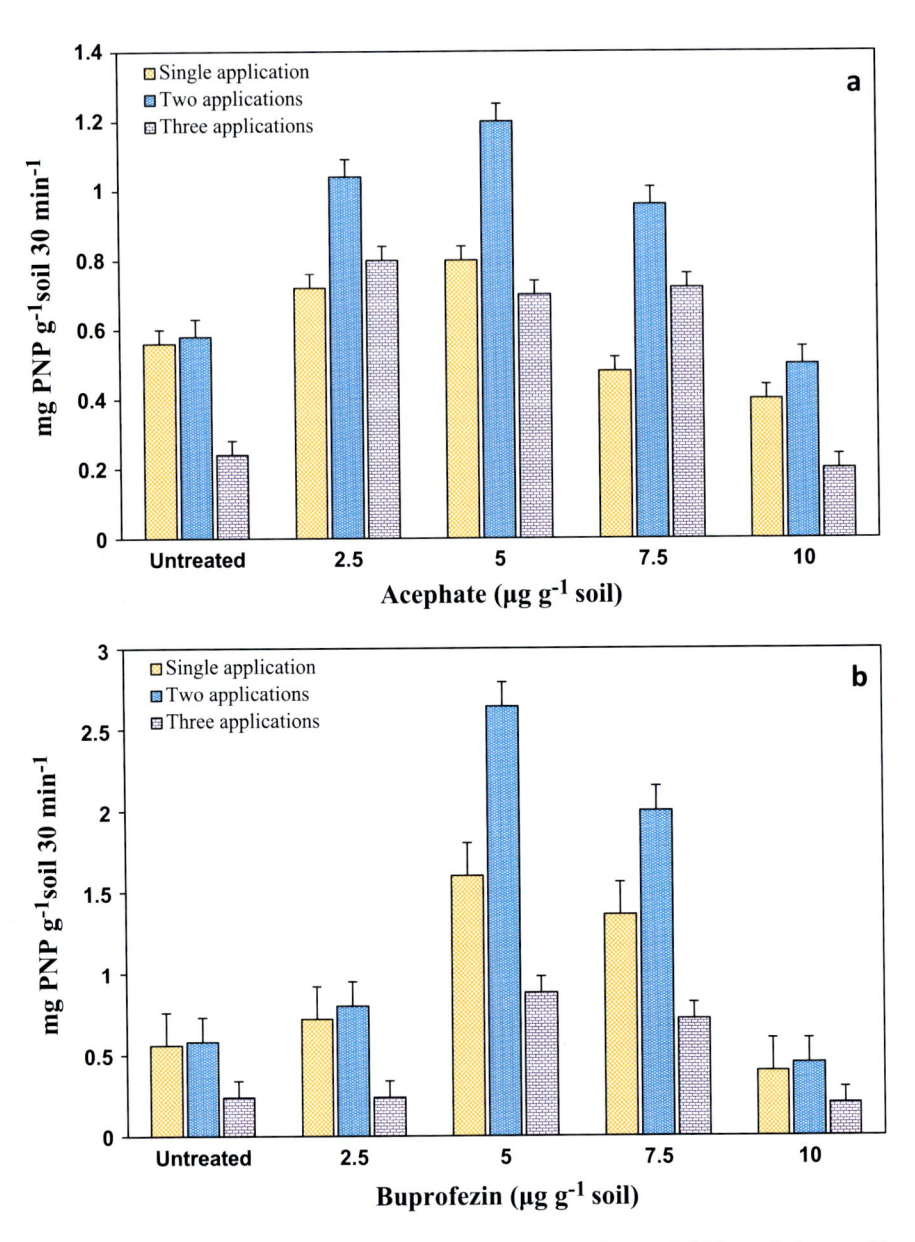

Fig. 9.1 Effect of single and repeated applications of (**a**) acephate and (**b**) buprofezin on acid phosphatase activity in soil. Error bars represent standard deviations ($n = 3$)

enzyme. But, the activity was stimulated after two applications of acephate to soil, and declined after three repeated applications.

Similarly, buprofezin stimulated (29–186%) the enzyme activity at concentrations ranging from 2.5 to 7.5 $\mu g\ g^{-1}$ (Fig. 9.1b). However, the insecticide was toxic (29% reduction in activity) to the enzyme at 10 $\mu g\ g^{-1}$ soil. The activity of the enzyme was stimulated (11–65%) even after two applications of buprofezin to soil. After three applications of buprofezin, the enzyme activity decreased (64–100%) drastically. The extent of acid phosphatase activity in NPK-amended soil samples under the impact of acephate and buprofezin was also determined and the results (Fig. 9.2) were very similar to those observed with soil samples that did not receive the fertilizers (Fig. 9.1). Both the insecticides significantly stimulated (up to 233%) the enzyme activity at 5 $\mu g\ g^{-1}$ soil. Nonetheless, the enzyme activity was inhibited greatly at higher rates of the two insecticides and, particularly, after three repeated applications.

Because of their significance in soil fertility, the changes of phosphatases activity in response to simultaneous and sequential applications of several pesticides were studied in laboratory and at field conditions. Omar and Abdel-Sater (2001) reported that brominal and selecron promoted acid phosphatase activity in soil at field application rates after some incubation periods, but the enzyme activity was less at higher rates of application. Sikora et al. (1990) reported that over 40% of the insecticide-treated soils had higher acid phosphatase activity than the fence row soils which had no previous exposure to chlorpyrifos, terbufos, fonofos, or phorate. Over 2/3rd of soils treated with fonofos had higher acid phosphatase and phosphotriesterase activity than the fence row soils. Stimulation in phosphatase activity under the influence of paraquat, trifluralin, glyphosate and atrazine was reported by Hazel and Greaves (1981). Four commonly used organophosphate insecticides such as monocrotophos, profenophos, quinalphos and triazophos at their field application rates of 0.75, 1.0, 0.5 and 0.6 kg a.i. ha^{-1}, respectively, increased acid phosphatase activity (Majumder and Das 2016).

The mean per cent values of increased acid phosphatase activity for monocrotophos, profenophos, quinalphos and triazophos were 21, 28, 43 and 24, respectively. However, fenamiphos (Megharaj et al. 1999) and flopet and captafol (Atlas et al. 1978) had no significant effects on phosphatases. Phosphatases are also highly sensitive to many other anthropogenic substances. For instance, the study of Sannino and Gianfreda (2001) reported a general inhibitory effect (5–98%) for phosphatase in the presence of glyphosate. Similarly, propiconazole, profenofos and pretilachlor (Kalam et al. 2004), acetamiprid (Yao et al. 2006) had a strong negative influence on phosphatase activity in soil. Soils treated with chlorothalonil (Yu et al. 2006), insecticides such as cyfluthrin, imidacloprid, tebupirimphos, aztec and amitraz (Tu 1995) showed short-term inhibitory effects. Pozo et al. (1995) suggested that the activities of acid phosphatase significantly decreased initially at concentrations of 2.0–10.0 kg ha^{-1} chloropyrifos, but recovered after 14 day to levels similar to those in control soil without the insecticide. On the other hand, phosphatases were shown to be stimulated in the beginning days of incubation and affected badly at later stages. Phosphatase activity in soils that received 2.5 kg ha^{-1} of dichlorovos, phorate

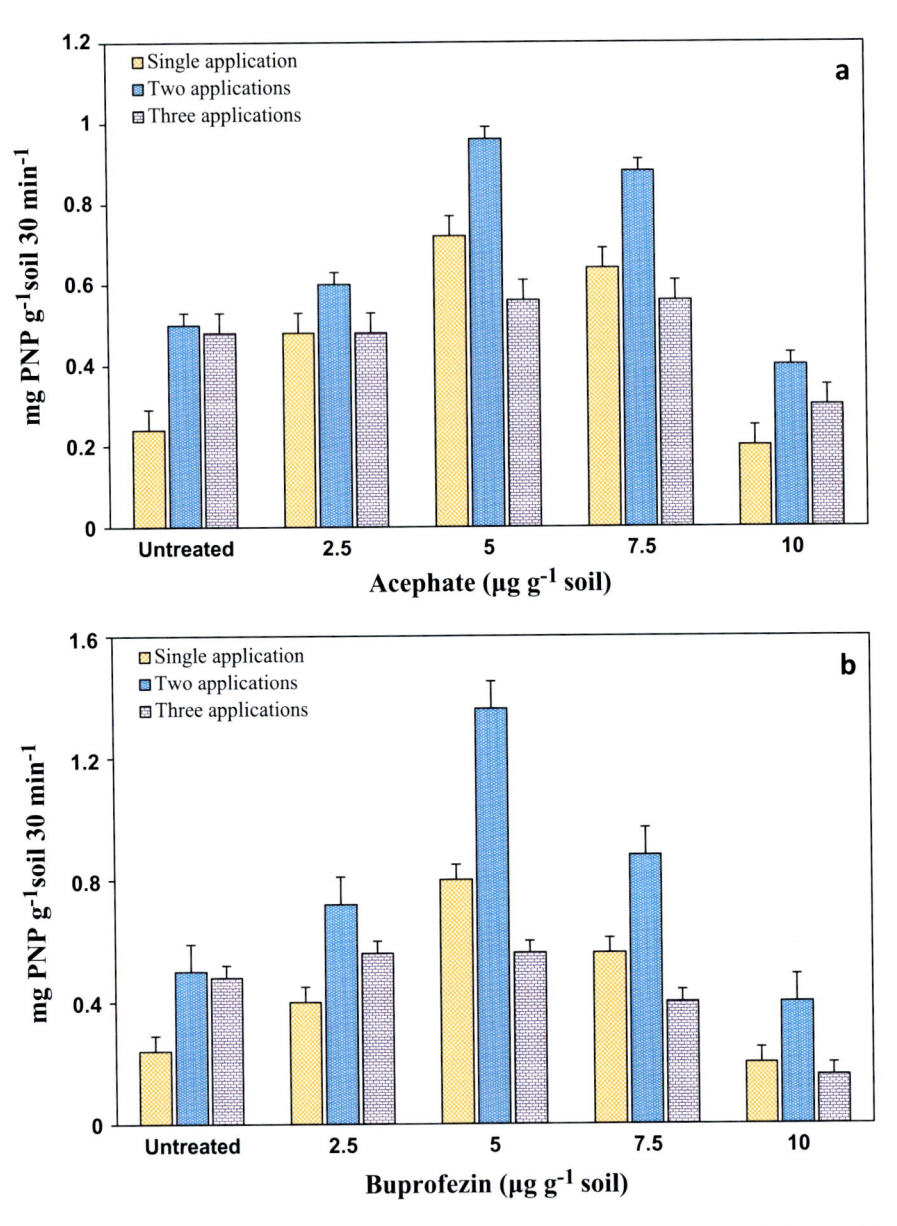

Fig. 9.2 Effect of single and repeated applications of (**a**) acephate and (**b**) buprofezin on acid phosphatase activity in soil amended with N-P-K. Error bars represent standard deviations (n = 3)

and methomyl, and 5.0 kg ha^{-1} of chloropyrifos and methyl parathion was significantly higher at 20 days, but further decreased progressively with incubation (Madhuri and Rangaswamy 2002). The activity of phosphatases in soils that received 10 and 100 ppm of endosulfan was significantly more at 42 days over untreated control, and further decreased progressively with incubation (Kalyani et al. 2010).

A widely used neonicotinoid insecticide, imidacloprid, was applied at two different concentrations (field rate of 1.0 mg kg^{-1} soil, and 10 mg kg^{-1} soil) to determine their effects on acid phosphatase in soil (Cycoń and Piotrowska-Seget 2015). In fact, imidacloprid at field rate had no significant impact on the enzyme after 14, 28 and 56 days, though the activity was slightly dropped (6% over untreated control) after 24 h. And, at 10 times the field rate of imidacloprid, the enzyme activity was decreased by 19 and 6% after day 1 and 14 days, respectively, but the inhibitory effect was nullified after 28 and 56 days. These results clearly indicate that higher doses of imidacloprid do not have any long-term adverse effects on acid phosphatases, even though the activity was negatively affected for a short-term. Very recently, Sanchez-Hernandez et al. (2017) reported that the activity of acid phosphatase decreased by 56–60% upon soil treatment with two doses (4.8 and 24 kg a.i. ha^{-1}) of chlorpyrifos. However, the response of acid phosphatase to chlorpyrifos was very similar to that with NaN$_3$ (Sodium azide)-soil (no microbial activity) and NaN$_3$ free-soil (with microbial activity) suspensions. Thus, these results not only confirm the sensitivity of acid phosphatases to chlorpyrifos but also provide information about the source (non-microbial origin) of acid phosphatase in soils.

Several other fungicides and insecticides also exhibited negative impact on acid phosphatases. Joseph et al. (2004) studied the effects of a fungicide, Ridomil Gold Plus copper, on soil chemical properties and microbial activities, and observed a slight decrease in phosphatase activity with high doses of the fungicide after 14 days of incubation. Voets et al. (1974) showed that long-term atrazine applications significantly reduced the activity of phosphatase in soil. Also, Jastrzebska and Kucharski (2007) reported that increasing doses of Swing Top 183 SC (a fungicide) significantly inhibited the activity of acid phosphatase. Chen et al. (2001) demonstrated that benomyl and captan inhibited the activity of acid phosphatase. Cypermethrin (Filimon et al. 2015) and mancozeb (Walia et al. 2014) decreased the enzyme activity by 13 and 7%, respectively. Surprisingly, Tarafdar (1986) observed a decrease in phosphatase activity due to fluchloralin, methabenzthiasuron, metoxuron, 2,4-D and isoproturon applied even at recommended field rates.

Application of fertilizers to soil minimizes the negative effects of pesticides on acid phosphatases in soil. For example, activity of acid phosphatase was lowered by 32 and 23% by dicamba and 2,4-D, respectively, when compared to that observed in soil without the pesticides (Oleszczuk et al. 2014). However, addition of biochar at a rate of 30 and 45 t ha^{-1} to soil significantly decreased the negative effects of 2,4-D and dicamba, respectively, toward acid phosphatases which indicated the alleviation mediated by biochar. The mechanism underlying the protective behavior of added nutrients against the negative effects of pesticides in soil may be one of several types, viz., sorption properties of fertilizers (as in case of biochar) (Sopeña and Bending, 2013; Tatarková et al. 2013), enrichment of

Table 9.1 Interaction effects of insecticide combinations on acid phosphatase activity* in soil

		Buprofezin (μg g^{-1} soil)				
		0	**2.5**	**5**	**7.5**	**10**
Acephate (μg g^{-1} soil)	**0**	Control	0.72 ± 0.08 28[a]	1.6 ± 0.08 186	1.36 ± 0.12 143	1.04 ± 0.04 86	
	2.5	0.72 ± 0.08 28	1.04 ± 0.2^C 86[a] 48[b]	1.44 ± 0.08^B 157 162	1.04 ± 0.04^C 186 131	0.88 ± 0.08^A 57 90	
	5	0.80 ± 0.01 43	0.91 ± 0.04^C 62 59	1.44 ± 0.12^B 157 149	0.40 ± 0.12^A −28 124	0.8 ± 0.08^A 43 92	
	7.5	0.48 ± 0.08 −14	0.96 ± 0.12^C 71 18	0.64 ± 0.16^A 14 198	0.64 ± 0.05^A 14 149	0.48 ± 0.12^A −14 84	
	10	0.40 ± 0.08 −28	0.56 ± 0.04^A 1 8	0.80 ± 0.08^A 43 210	1.04 ± 0.2^A 86 155	0.96 ± 0.12^A 71 82	

*mg PNP g^{-1} 30 min^{-1}

Control value, 0.56 ± 0.08 mg PNP g^{-1} 30 min^{-1}

All entries are means (n = 3) of per cent stimulation/inhibition values of enzyme activity relative to untreated control

[a]Experimental per cent values (first row) over control

[b]Expected per cent values (second row) over control

A: *Antagonistic* insecticide interaction

B: *Additive* insecticide interaction

C: *Synergistic* insecticide interaction

specific taxa within the microbial community by added fertilizers (Sopeña and Bending, 2013), or sometime there would be a promotion of redox reactions by added fertilizers, leading to reductive transformation of pesticides (Oh et al. 2013). All such mechanisms lead to the decrease in the concentration of pesticides in the environment.

The response of acid phosphatase in the presence of acephate or buprofezin, in combination, at graded concentrations was studied in soil samples with and without NPK-fertilizer amendments. In fact, the enzyme was greatly affected in unamended soil (Table 9.1) than in fertilizer-amended soil (Table 9.2). Buprofezin at 2.5 μg g^{-1} soil level interacted synergistically with acid phosphatase when it was combined

Table 9.2 Interaction effects of insecticide combinations on acid phosphatase activity* in soil amended with N-P-K

		Buprofezin (µg g⁻¹ soil)				
		0	**2.5**	**5**	**7.5**	**10**
Acephate (µg g⁻¹ soil)	**0**	Control	0.4 ± 0.16 67[a]	0.8 ± 0.04 233	0.56 ± 0.04 133	0.48 ± 0.08 100
	2.5	0.48 ± 0.04 100	0.72 ± 0.04[C] 200[a] 100[b]	0.56 ± 0.04[C] 133 100	0.52 ± 0.04[C] 117 100	0.4 ± 0.04[A] 67 100
	5	0.72 ± 0.12 200	0.64 ± 0.04[C] 300 133	0.08 ± 0.24[C] −50 −33	1.36 ± 0.12[C] 467 67	0.32 ± 0.12[A] 33 100
	7.5	0.64 ± 0.04 167	0.8 ± 0.04[C] 233 122	0.56 ± 0.04[C] 133 11	0.32 ± 0.08[A] 33 78	0.4 ± 0.04[A] 67 167
	10	0.56 ± 0.04 133	0.48 ± 0.04[A] 100 111	0.16 ± 0.08[A] −33 56	0.32 ± 0.04[A] 33 89	0.32 ± 0.04[A] 33 100

*mg PNP g⁻¹ 30 min⁻¹
Control value, 0.24 ± 0.04 mg PNP g⁻¹ 30 min⁻¹
All entries are means (n = 3) of per cent stimulation/inhibition values of enzyme activity relative to untreated control
[a]Experimental per cent values (first row) over control
[b]Expected per cent values(second row) over control
A: *Antagonistic* insecticide interaction
B: *Additive* insecticide interaction
C: *Synergistic* insecticide interaction

with acephate at concentrations ranging from 2.5–7.5 µg g⁻¹ soil (Table 9.1). Again, 5 µg g⁻¹ soil levels of buprofezin in combination with acephate at 2.5 and 5 µg g⁻¹ yielded additive responses. However, the combination at higher rates of insecticides resulted in antagonistic interaction with soil acid phosphatase. On the other hand, buprofezin treatment up to 7.5 µg g⁻¹ interacted synergistically with acid phosphatase in NPK-amended soil when it was combined with acephate even up to the level of 7.5 µg g⁻¹ soil (Table 9.2).

The negative effects of insecticide combination were thus alleviated in the presence of nutrient fertilizer. But, the interaction was antagonistic when the two insecticides were used in combination at higher levels. The above result clearly indicates that the insecticides when applied at concentrations close to field application rates stimulate the activity of acid phosphatase in soil.

Table 9.3 Interaction effects of insecticide combinations on alkaline phosphatase activity* in soil

		Buprofezin (μg g^{-1} soil)				
		0	**2.5**	**5**	**7.5**	**10**
Acephate (μg g^{-1} soil)	**0**	Control	0.96 ± 0.01 1[a]	1.12 ± 0.02 17	2.24 ± 0.01 273	0.88 ± 0.03 −8
	2.5	1.12 ± 0.02 17	1.2 ± 0.03[C] 25[a] 18[b]	1.28 ± 0.03[C] 33 31	0.8 ± 0.02[A] −17 243	0.48 ± 0.01[A] −50 10
	5	0.64 ± 0.02 −33	0.4 ± 0.01[C] −58 −32	0.88 ± 0.01[A] −8 −11	1.76 ± 0.03[A] 83 331	0.8 ± 0.01[A] −17 44
	7.5	0.64 ± 0.03 −33	0.5 ± 0.01[C] −48 −32	0.8 ± 0.02[C] −17 −11	0.64 ± 0.02[A] −33 331	0.32 ± 0.02[A] −47 44
	10	0.4 ± 0.02 −58	0.72 ± 0.02[A] −25 −56	0.7 ± 0.01[A] −27 −31	0.24 ± 0.02[A] −75 373	0.16 ± 0.01[A] −83 71

*mg PNP g^{-1} 30 min^{-1}

Control value, 0.96 ± 0.03 mg PNP g^{-1} 30 min^{-1}

All entries are means (n = 3) of per cent stimulation/inhibition values of enzyme activity relative to untreated control

[a]Experimental per cent values (first row) over control.

[b]Expected per cent values (second row) over control

A: *Antagonistic* insecticide interaction

B: *Additive* insecticide interaction

C: *Synergistic* insecticide interaction

Table 9.4 Interaction effects of insecticide combinations on alkaline phosphatase activity* in soil amended with N-P-K

		Buprofezin (µg g⁻¹ soil)				
		0	**2.5**	**5**	**7.5**	**10**
Acephate (µg g⁻¹ soil)	**0**	Control	0.08 ± 0.01 −83[a]	0.24 ± 0.01 −50	0.4 ± 0.01 −17	0.4 ± 0.01 −17
	2.5	0.56 ± 0.02 17	0.16 ± 0.01[C] −67[a] −53[b]	0.24 ± 0.02[C] −50 −25	0.32 ± 0.01[A] −33 3	0.16 ± 0.01[A] −67 3
	5	0.24 ± 0.03 −50	0.4 ± 0.02[A] −17 153	0.16 ± 0.01[A] −67 −125	0.24 ± 0.01[A] −50 75	0.08 ± 0.01[A] −83 75
	7.5	0.08 ± 0.01 −83	0.08 ± 0.01[A] −83 236	0.56 ± 0.01[A] 17 −175	0.24 ± 0.01[A] −50 144	0.08 ± 0.01[A] −83 114
	10	0.08 ± 0.01 −83	0.32 ± 0.01[A] −33 236	0.48 ± 0.03[A] 1 −175	0.08 ± 0.01[A] −83 −114	0.08 ± 0.01[A] −83 114

*mg PNP g⁻¹ 30 min⁻¹

Control value, 0.48 ± 0.01 mg PNP g⁻¹ 30 min⁻¹

All entries are means (n = 3) of per cent stimulation/inhibition values of enzyme activity relative to untreated control

[a]Experimental per cent values (first row) over control

[b]Expected per cent values (second row) over control

A: *Antagonistic* insecticide interaction

B: *Additive* insecticide interaction

C: *Synergistic* insecticide interaction

Nontarget Effects on Alkaline Phosphatase

The data on interaction effects resulting from combination of the two insecticides at graded concentrations toward alkaline phosphatase activity in soil amended with or without NPK-fertilizer are presented in Tables 9.3 and 9.4. Interaction effects of the two insecticides turned out to be detrimental to the enzyme in NPK-amended soil than in unamended soil. Buprofezin either at 2.5 or 5 µg g⁻¹ soil level interacted synergistically with the enzyme when it was combined with acephate up to 7.5 µg g⁻¹ soil (Table 9.3). However, the interaction of the two insecticides in combination at higher rates was antagonistic to the enzyme activity. Acephate treatment, on the other hand, at 2.5 μg g^{-1} soil in combination with buprofezin up to 5 μg g^{-1} soil level in NPK-amended soil resulted in synergistic interaction with the alkaline phosphatase activity (Table 9.4). The two higher concentrations included (7.5 and 10 μg g^{-1} soil) exhibited exclusively antagonistic response on alkaline phosphatase. Even the concentration of 5 μg g^{-1} soil of acephate interacted antagonistically with the enzyme when combined with buprofezin at different levels. The above results clearly suggest that alkaline phosphatase in soil is very sensitive to insecticides, particularly with NPK-amendments.

In many other investigations also most pesticides exerted adverse effects on alkaline phosphatases in soil. Alkaline soils treated with brominal and selecron exhibited increased alkaline phosphatase activity even at higher applications rates (Omar and Abdel-Sater 2001). Similarly, there was stimulation in the activity of alkaline phosphatase by four OP insecticides at the field level application rates (Majumder and Das 2016). The mean values of increased enzyme activities by monocrotophos, profenophos, quinalphos and triazophos were 164, 152, 136 and 135, respectively, over untreated controls (131 μg PNP g^{-1} h^{-1}). Some fungicides like Unix 75 WG also stimulated alkaline phosphatases in soil (Jastrzebska and Kucharski 2007).

On the other hand, most of other pesticides studied affected the enzyme activities negatively. For example, Cervelli et al. (1978) observed that all vinyl phosphate insecticides investigated (viz., dichlorovos, tetrachlorvinphos, crotoxyphos and phosphomidon) were competitive inhibitors of alkaline phosphatase. While investigating the impact of imidacloprid at a field application rate (1 mg kg^{-1} soil) on alkaline phosphatase in soil, Cycoń and Piotrowska-Seget (2015) observed that there was a marginal decrease (3–5%) in the enzyme activity in the first 2 weeks, and remained unaffected after that (4–8 weeks). However, at higher doses of imidacloprid (10 times the field application rate), the enzyme activity was negatively affected (4–16%) throughout. It was also found that the alkaline phosphatase was more sensitive to imidacloprid than acid phosphatase. In fact, alkaline phosphatases are not only sensitive to various xenobiotics at different levels, but also to several metals ranging from 7.2 to 48.1 mmol kg^{-1} (Kuperman and Carreiro 1997). A short-term inhibitory effect of alkaline phosphatase activity was noticed in soil treated with chlorothalonil (Yu et al. 2006). Similar observation was reported by Tu (1995) in soil treated with 5 different insecticides. Kalam et al. (2004) observed nearly 46% inhibition of phosphatase activity in soil that received 100 mg kg^{-1} propiconazole after 120 days. Yao et al. (2006) found strong negative influence on phosphatase activity in soil treated with a new neonicotinoid insecticide, acetamiprid, applied at normal field concentration (0.5 mg kg^{-1}) as well as high concentrations (5 and 50 mg kg^{-1}). Likewise, phosphatase activities were adversely affected when soils were treated with glyphosate (Sannino and Gianfreda 2001), and propiconazole (Kalam et al. 2004). Several fungicides have also been investigated for their effects on activities of alkaline phosphatase in soil. For instance, Swing Top 183 SC showed strong negative effects on enzyme, especially when applied at higher rates (Jastrzebska and Kucharski 2007). In particular, alkaline phosphatase was found to be the most sensitive enzyme to fungicides in soils (Monkiedje et al. 2002). Two other fungicides namely tebuconazole (TEB) and carbendazim (CAB) were tested for their effects when applied either alone or in combination on activities of alkaline phosphatase in soil (Wang et al. 2016). Although individual fungicides did not exert any effect at 1 mg kg^{-1}, mixture of the two fungicides significantly inhibited the enzyme activity. Carbendazim was not inhibitory for the enzyme even at 10 mg kg^{-1}. On the other hand, tebuconazole alone or in combination with carbendazim at 10 mg kg^{-1} significantly lowered the activities of alkaline phosphatase throughout the study period of 90 days (Wang et al. 2016). Nevertheless, higher rates (i.e., 100 mg kg^{-1}) of fungicides (individual or mixture), caused adverse effects on enzyme activities in the first 60 days, and the order of inhibition was TEB and CAB > TEB > CAB.

In support of results presented in Table 9.4, there are some reports where negative effects of pesticides were neutralized by other added organic substances. For instances, lower doses of dicamba and 2,4-D increased the activity of alkaline phosphatase by 76 and 104%, respectively (Oleszczuk et al. 2014). But higher doses of those chemicals have adversely affected the enzyme activity. However, such negative effects of 2,4-D on alkaline phosphatase were alleviated in soil by the addition of biochar (45 t ha^{-1}). In all, the present data demonstrate that the addition of nutrient fertilizers could considerably alter the toxic effects of insecticides, when applied alone, toward soil alkaline phosphatase activity.

References

Atlas RM, Pramer D, Bartha R (1978) Assessment of pesticide effects on non-target soil microorganisms. Soil Biol Biochem 10:231–239

Cervelli S, Nannipieri P, Giovannini G, Pernal A (1978) Alkaline phosphatase inhibition by vinyl phosphate insecticides. Water Air Soil Pollut 9:315–321

Chen SK, Edwards CA, Subler S (2001) Effects of the fungicides benomyl, captan and chlorothalonil on soil microbial activity and nitrogen dynamics in laboratory incubations. Soil Biol Biochem 33:1971–1980

Cycoń M, Piotrowska-Seget Z (2015) Biochemical and microbial soil functioning after application of the insecticide imidacloprid. J Environ Sci 27:147–158

Eivazi F, Tabatabai MA (1977) Phosphatase in soil. Soil Biol Biochem 9:167–172

Filimon MN, Voia SO, Popescu R, Dumitrescu G, Ciochina LP, Mituletu M, Vlad DC (2015) The effect of some insecticides on soil microorganisms based on enzymatic and bacteriological analyses. Rom Biotech Lett 20:10439–10447

Hazel AD, Greaves MP (1981) Effects of some herbicides on soil enzyme activities. Weed Res 21:205–209

Jastrzebska E, Kucharski J (2007) Dehydrogenases, urease and phosphatases activities of soil contaminated with fungicides. Plant Soil Environ 53:51–57

Joseph D, Adolphe M, Thomas N, Samuel MF, Moise N, Serges HZT, Norbert K (2004) Changes in soil chemical properties and microbial activities in response to the fungicide Ridomil gold plus copper. Int J Environ Res Pub Health 1:26–34

Kalam A, Tah J, Mukherjee AK (2004) Pesticide effects on microbial population and soil enzyme activities during vermicomposting of agricultural waste. J Environ Biol 25:201–208

Kalyani SS, Sharma J, Dureja P, Singh S, Lata (2010) Influence of endosulfan on microbial biomass and soil enzymatic activities of a tropical alfisol. Bull Environ Contam Toxicol 84:351–356

Kuperman RG, Carreiro MM (1997) Soil heavy metal concentrations microbial biomass and enzyme activities in a contaminated grassland ecosystem. Soil Biol Biochem 29:179–190

Madhuri RJ, Rangaswamy V (2002) Influence of selected insecticides on phosphatase activity in groundnut (*Arachis hypegeae* L.) soils. J Environ Biol 23:393–397

Majumder SP, Das AC (2016) Phosphate-solubility and phosphatase activity in Gangetic alluvial soil as influenced by organophosphate insecticide residues. Ecotoxicol Environ Saf 126:56–61

Megharaj M, Singleton I, Kookana R, Naidu R (1999) Persistence and effects of fenamiphos to native algal populations and enzyme activities in soil. Soil Biol Biochem 31:1549–1553

Monkiedje A, Ilori MO, Spiteller M (2002) Soil quality changes resulting from the application of the fungicides mefenxomam and metalaxyl to a sandy loam soil. Soil Biol Biochem 18:31–37

Oh SY, Son JG, Chiu PC (2013) Biochar-mediated reductive transformation of nitro herbicides and explosives. Environ Toxicol Chem 32:501–508

Oleszczuk P, Jośko I, Futa B, Pasieczna-Patkowska S, Pałys E, Kraska P (2014) Effect of pesticides on microorganisms, enzymatic activity and plant in biochar-amended soil. Geoderma 214–215:10–18

Omar SA, Abdel-Sater MA (2001) Microbial populations and enzyme activities in soil treated with pesticides. Water Air Soil Pollut 127:49–63

Pozo C, Martinez-Toledo MV, Salmeron V, Rodelas B, Gonzales-Lopez J (1995) Effect of chlorpyrifos on soil microbial activity. Environ Toxicol Chem 14:187–192

Sanchez-Hernandez JC, Sandoval M, Pierart A (2017) Short-term response of soil enzyme activities in a chlorpyrifos-treated mesocosm: use of enzyme-based indexes. Ecol Indic 73:525–535

Sannino F, Gianfreda L (2001) Pesticide influence on soil enzymatic activities. Chemosphere 45:417–425

Sikora LJ, Kaufman DD, Horng LC (1990) Enzyme activity in soils showing enhanced degradation of organophosphate insecticides. Biol Fertil Soils 9:14–18

Sopeña F, Bending GD (2013) Impacts of biochar on bioavailability of the fungicide azoxystrobin: a comparison of the effect on biodegradation rate and toxicity to the fungal community. Chemosphere 9:1525–1533

Tabatabai MA, Bremner JM (1969) Use of *p*-nitrophenyl phosphate for assay of soil phosphatase activity. Soil Biol Biochem 1:301–307

Tarafdar JC (1986) Effect of different herbicides on enzyme activity in controlling weeds in wheat. Pesticides 20:46–49

Tatarková V, Hiller E, Vaculík M (2013) Impact of wheat straw biochar addition to soil on the sorption, leaching, dissipation of the herbicide (4-chloro-2-methylphenoxy) acetic acid and the growth of sunflower (*Helianthus annuus* L.) Ecotoxicol Environ Saf 92:215–221

Tu CM (1995) Effect of five insecticides on microbial and enzymatic activities in sandy soil. J Environ Sci Health B30:289–306

Voets JP, Meerschman P, Verstraete W (1974) Soil microbiological and biochemical effects of long-term atrazine applications. Soil Biol Biochem 6:149–152

Walia A, Mehta P, Guleria S, Chauhan A, Shirkot CK (2014) Impact of fungicide mancozeb at different application rates on soil microbial populations, soil biological processes, and enzyme activities in soil. Sci World J 2014:1–9

Wang C, Wang F, Zhang Q, Liang W (2016) Individual and combined effects of tebuconazole and carbendazim on soil microbial activity. Eur J Soil Biol 72:6–13

Yao X, Min H, Lu Z, Yuan H (2006) Influence of acetamiprid on soil enzymatic activities and respiration. Eur J Soil Biol 42:120–126

Yu YL, Shan M, Fang H, Wang X, Chu XO (2006) Responses of soil microorganisms and enzymes to repeated applications of chlorothalonil. J Agric Food Chem 54:10070–10075

Chapter 10
Bacterial Utilization of Acephate and Buprofezin

Introduction

OP pesticides have been extensively applied as alternatives to organochlorine compounds which are long-term persistent and highly toxic. In general, OP compounds rapidly undergo degradation by soil microorganisms, so they do not persist in the environment. However, repeated applications of degradable organophosphates occasionally cause a significant reduction of their pesticidal effect. This phenomenon which results from microbial adaptation to pesticide degradation, called enhanced biodegradation, has often been observed in degradable pesticides such as organophosphates and carbamates. Most enhanced degradation in the field occurs after pesticide applications for two or more consecutive years (Racke and Coats 1990). Many of these insecticides are phosphorous thioesters, with limited aqueous solubility, in which the leaving group is attached to the phosphorous center via a sulfur atom, such as malathion, dementon-s, acephate, azinophos-ethyl, and phosalone.

Native environmental biodegradation has been observed to have the potential to provide an effective means of detoxifying modest levels of environmental pollutants (Landis and Frank 1991; Harkness et al. 1993). For example, soil samples selected from contaminated environments have been reported to degrade various OP-thioates in the field following repeated application of these insecticides (Chapalamadugu and Chaudhry 1992). In recent years, OP compounds have been the most widely used group of insecticides in India which include acephate, endosulfan, phosphomidon, and dimethoate. Approximately 35 tones of OP-active ingredients comprising about 4 distinct compounds have been used annually in Nandyal region, Andhra Pradesh, India for many years. On the other hand, buprofezin, a thiadiazine compound, is highly effective in controlling harmful insect pests including rice brown planthopper (*Nilaparvata lugens*) and greenhouse whitefly (*Trialeurodes vaporariorum*). This insecticide is being used at a level >30,000 L y^{-1} in Nandyal division alone on various crops including the two major crops, rice and cotton.

N.R. Maddela, K. Venkateswarlu, *Insecticides−Soil Microbiota Interactions*,
DOI 10.1007/978-3-319-66589-4_10

Utilization of OP Insecticides by Soil Bacteria

OP compounds are among the most common chemical classes used in crop and live-stock protection and account for an estimated 34% of world-wide insecticide scales. OP compounds possess very high mammalian toxicity and therefore early detection and subsequent decontamination and detoxification of the polluted environment is essential. The wide use of OP pesticides such as isofenphos, chlorpyrifos, diazinon, phorate, ethoprophos, terbufos, phosalone, pirimphos methyl has created numerous problems, including pollution of the environment. OP pesticides, in general, are regarded as non-persistent. Chemical and physical methods of decontamination are not only expensive and time-consuming, but also in most cases they do not provide a complete solution. These approaches convert toxic compounds into less toxic states, which in some cases can accumulate in the environment and still to be toxic to a range of organisms. We now know that bioremediation provides a suitable way to remove contaminants from the environment as, in most of the cases, OP compounds are totally mineralized by the microorganisms. Fortunately, most OP compounds are degraded by microorganisms in the environment as a source of phosphorus or carbon or both. In this direction, several soil bacteria have been isolated and characterized, which can degrade OP compounds in laboratory cultures and in the field. Likewise, the bio-chemical and genetic basis of microbial degradation has received considerable atten-tion. Available literature on the microbial degradation of xenobiotics indicates that most studies have considered three aspects: (i) the fundamental basis of biodegrada-tion, (ii) evolution and transfer of such activities among microorganisms, and (iii) bioremediation techniques to detoxify contaminated environment (Singh et al. 1999). However, the use of microorganisms for bioremediation requires an understanding of all physiological, microbiological, ecological, biochemical and molecular aspects involved in pollutant transformation (Iranzo et al. 2001). The net result of interaction between xenobiotics and soil microflora is notoriously difficult to predict. Because, microbial communities that can degrade or can develop tolerance to, or are inhibited by, chemical mixtures greatly contribute to resilience and resistance in soil environ-ments (Ramakrishnan et al. 2011).

In recent years, the role of soil microorganisms in affecting the persistence of agricul-tural pesticides has been the subject of two areas of study. The first is the capacity for rapid elimination of highly persistent or toxic chemicals. The reduced pesticide efficacy is attributed to enhanced biodegradation particularly of chemicals applied under a con-tinuous cropping program. In one study, a streptomycete was isolated from a field soil sample previously treated with the insecticide isofenphos and found to be capable of growing on several commercial carbamates and OP insecticides (Gauger et al. 1986). In another laboratory study, degradation of widely used OP insecticide, monocrotophos in two Indian agricultural soils at two concentrations, 10 and 100 µg g^{-1} soil, under aerobic conditions (60% water-holding capacity) at 28 ± 4 °C was studied by Gundi and Reddy (2006). The degradation of monocrotophos at both concentrations in black vertisol and red alfinsol was rapid accounting for 96–98% disappearance of the applied chemical and followed the first-order kinetics. The rate constants (k) for vertisol and alfinsol were 0.0753 and 0.0606 day^{-1}, and half lives were 9.2 and 11.4 days, respectively.

Catabolism and detoxification occur when a soil microorganisms uses the pesticide as a carbon and energy source. The later process is facilitated by resistant microorganisms (Matsumura 1988). The reduced persistence of OP insecticides was attributed to the activity of soil microorganisms (Chapman and Harris 1982; Gorder et al. 1982; Sharmila et al. 1989). The degradation of xenobiotic compounds by members of soil microflora is an important means by which these compounds are removed from the environment, thus preventing them from becoming pollution problems. Much work has been directed towards understanding the complexity of pesticide–microflora interactions in soil. Many studies have employed pure cultures of soil isolates or agar plate counts of soil populations (Visalakshi et al. 1980; Digrak and Ozcelik 1998). Racke and Coats (1988) reported that a bacterial strain (*Pseudomonas*) was isolated from an isofenphos-treated culture medium, and it proved capable of using isofenphos as a carbon source. Several *Pseudomonas* spp. that metabolize OP and carbamate insecticides have been isolated from soil (Siddaramappa et al. 1973; Chaudhry and Wheeler 1988). Some OP insecticides, such as diazinon, chlorpyrifos, ethion, parathion, fonofos, malathion and gusathion, are susceptible to microbial hydrolysis and many serve as carbon sources for the growth of pure and mixed cultures *Flavobacterium* sp., *Pseudomonas* sp., and *Arthrobacter* sp. (Ghisalba et al.1987; Digrak et al.1995). Gauger et al. (1986) reported that *Streptomyces pilosus* was capable of growing on several insecticides (carbofuran, cloethocarb, trimethacarb isofenphos, fonofos and phorate) although growth on terbufos was found to be nonexistent.

Digrak and Kazanici (2001) reported that the total viable bacterial count in the isofenphos-treated soil sample was found to be higher than that of the untreated control soil samples during incubation. Moreover, it was observed that the treatment had no inhibitory effect on the development of other groups of microorganisms. Also, isofenphos-degrading *Arthrobacter* sp. was able to rapidly metabolize this compound. A granular formulation (5%) of monocrotophos, applied at a rate of 1.5 g a.i. ha^{-1} to an Indian clay soil, was dissipated rapidly with a half-life of 10 days (Agnihotri et al. 1981). Enhanced biodegradation responsible for rapid loss of another OP insecticide, chlorpyrifos from Australian cane fields was attributed to fallen efficacy against cané grub (Robertson et al. 1998). Repeated treatments with an OP nematicide, fenamiphos resulted in enhanced biodegradation of the compound in soils of the United Kingdom with high pH (7.7) but not in soils with acidic pH (Singh et al. 2003). Adebayo et al. (2007) reported that the bacterial population in soil was significantly increased upon soil treatment with karate and thiodan. Among the pesticides, few significant effects of herbicides on soil organisms have been documented by Bunemann et al. (2006). Such reports are important because they disclose the behavior and potential harm caused by these chemicals in the environment, and these reports are useful in identifying data gaps for remediation by future research.

Acephate, one of the important OP foliar spray insecticide, is used for control of a wide range of biting and sucking insects. This insecticide dissipates rapidly with half-lives of <3 and 14 days in aerobic and anaerobic soils, respectively. Laboratory degradation studies have been demonstrated that acephate can degrade through microbial degradation and aqueous hydrolysis. Since the rate of hydrolysis increases with increasing pH, degradation may occur more rapidly in alkaline soil than in acidic soil. A review of available literature indicated that the average half-life for acephate is 3–4 days under aerobic (flooded) conditions (Chevron 1972a). Furthermore, acephate is rapidly degraded in soil by microorganisms under both aerobic and anaerobic conditions. The soil types in this experiment included loamy sand, sandy clay, silty clay loam, loam, and clay (Chevron 1972a, b). The same degradation products are formed in both aerobic and anaerobic soils. The metabolites formed are methamidophos and O-methyl N-acetylphosphoramidate (Chevron 1972c, d). According to US EPA (1987), acephate dissipates rapidly with half-lives of less than 3 and 6 days in aerobic and anaerobic soils, respectively. The major metabolite was found to be carbon dioxide in both types of soils. Zhi et al. (2008) provided a thorough review of the literature that presented acephate-degrading bacteria isolated from soils in which acephate was used for a long period. The bacterial strain that belongs to *Chrysobacterium* sp. XP-3 has a strong ability of growth and reproduction in medium containing 1500 mg L^{-1} acephate.

Major advances in the degradation of buprofezin, an important insecticide, in flooded and upland soils under laboratory conditions have been made by Funayama et al. (1986). Buprofezin was gradually decomposed in soils under flooded and upland conditions, with half-lives of 104 and 80 days, respectively. After 150 days, five degradation products were identified by thin-layer cochromatography which include: 2-*tert*-butylimnio-5-(4-hydroxyphenyl)-3-isopropyl-perhydro-1,3,5-thiadiadin-4-one, 3-isopropyl-5-phenyl-perhydro-1,3,5-thiadiazin-2, 4-dione, 1-*tert*-butyl-3-isopropyl-5-phenyl-biuret, 1-isopropyl-3-phenylurea and phenylurea. As minor products, 2-*tert*-burylimino-5-phenyl-perhydro-1,3,5-thiaxiazin-4-one or buprofezin sulfoxide were found in the flooded or the upland soil. [^{14}C] Carbon dioxide and bound ^{14}C residue accounted for 23–24% and 13–21% of the applied radioactivity, respectively. Degradation of buprofezin remarkably delayed in sterile soils. Since neither formation of $^{14}CO_2$ nor ring hydroxylation was observed in the sterile soils, buprofezin seems to have undergone complete mineralization in the soils under both flooded and upland conditions through biological transformation by soil microorganisms. Though biodegradation of OP insecticides and other pesticides by microorganisms in soil has been widely reported (Racke and Coats 1988; Sharmila et al. 1989; Digrak 1994; Digrak et al. 1995), the impact of soil bacteria on acephate and buprofezin has received less attention.

Selective Enrichment of Bacteria with Acephate and Buprofezin

Often, it is desired to isolate bacteria that are relatively scarce, or are in fact in very low numbers following the basic principle of selection. In the present study, samples (20 g, fresh weight) of the same soil, used for investigating the nontarget effects of the selected insecticides, were treated with acephate or buprofezin repeatedly at

25 μg g^{-1} active ingredient for six consecutive times at an interval of 7 days. The moisture content was maintained at 60% water-holding capacity throughout the experimental period by checking gravimetrically once in every 2 days. A week after 6th addition, 1 g of soil was taken, subjected it to serial dilution. An aliquot of 0.1 mL diluted soil sample was placed on nutrient agar medium, distributed uniformly with a sterile spreader under aseptic conditions. Three replicate plates were incubated at 37 °C overnight in an inverted position. Single colonies growing on the medium were tested for their ability to utilize acephate or buprofezin as energy and/ or elemental source. Bacterium that appeared predominantly from the enrichment cultures was preserved in the form of glycerol stocks till their use.

Bacterial Growth Experiments

Minimal salts medium (MSM) of pH 7 contained (g L^{-1}): K_2HPO_4, 2.27; $Na_2HPO_4.12H_2O$, 5.97; $MgSO_4.7H_2O$, 0.5; $CaCl_2.2H_2O$, 0.1; $MnSO_4.4H_2O$, 0.02; $FeSO_4$, 0.005 (Roberts et al.1993). Initially, 50 mL portions of MSM contained in a 100 mL of Erlenmeyer flask supplemented with 25 μg g^{-1} acephate or buprofezin and 1.25 g^{-1} of glucose were inoculated with a single colony of the isolated bacterium. The culture was incubated in a shaker at 37 °C till the culture density reached to ~0.7 A_{600}. The culture medium was then centrifuged at 2500 rpm for 20 min to obtain a bacterial cell pellet, and the pellet was washed twice in distilled water, suspended in 5 mL of MSM and was used further as an inoculum to give a final density (A_{600}) of 0.2. Finally, the culture flask was kept in a shaker at 37 °C. At regular intervals up to 48 h, 50 μL of culture broth was withdrawn aseptically to determine the viable cell density of the bacterium by spreading on nutrient agar medium. After incubation at 37 °C overnight, the number of bacterial colonies was recorded using a colony counter. Culture grown in MSM containing no insecticide served as control. Toxicity of the selected insecticides towards growth of the isolated bacterium was studied by including different concentrations (50, 100, 200 and 500 μg mL^{-1}) of acephate or buprofezin in MSM. The influence of different carbon sources (0.01%) such as glucose, yeast extract, and sodium acetate on growth of the bacterium in MSM supplemented with acephate or buprofezin at 50 μg mL^{-1} was also investigated. Using MSM containing 0.01% sodium acetate as a carbon source, growth of the bacterial isolate was assessed in presence of 0.15% of NH_4Cl, NH_4NO_3, or NH_6PO_4. Since NH_4Cl appeared to be a better source of nitrogen in promoting growth of the bacterium in presence of the insecticides, concentrations of 0.075%, 0.15%, and 0.3% of this ammonical source were included to check for toxicity of acephate or buprofezin in MSM. Growth of the bacterial isolate was also assessed on minimal salts agar (MSA) medium supplemented with acephate or buprofezin. MSA contained MSM and Difco Bacto agar (15 g L^{-1}). In order to remove water-soluble contaminants, Difco Bacto agar was washed twice with distilled water. The extent of growth of bacterial colonies with time on agar medium was observed.

Growth of *Pseudomonas* sp. in Presence of Acephate and Buprofezin

The bacterial strain, isolated from soil following selective enrichment using acephate, was a Gram negative rod. Based on several biochemical characteristics such as gelatin hydrolysis, catalase, carbohydrate fermentation, indole production, methyl red, Voges-Proskauer, citrate utilization, urease, nitrate reduction, starch hydrolysis (Table 10.1) and by referring to the Bergey's Manual, the bacterium has been tentatively identified as a species of *Pseudomonas*.

The potential of the bacterial strain in utilizing acephate or buprofezin as sole source of carbon and/or energy was assessed by determining growth (A_{600}) after 24 h incubation in MSM that contains no carbon source, supplemented with an insecticide. Acephate has been utilized by the bacterium in MSM supplemented with 25 µg mL^{-1} as a sole carbon and nitrogen source, and maximum growth (220%) as evidenced by culture turbidity was observed after 6 h of incubation. The viable cell number was determined after plating an aliquot the culture medium. The cell number in MSM containing acephate at the start of the experiment was 8×10^3 CFU mL^{-1}. After 2, 4 and 6 h, bacterial cell number in the culture medium was 14.4×10^3, 20×10^3, and 25.6×10^3 CFU mL^{-1}, respectively. There was no change in bacterial cell density afterwards.

Bacterial growth was apparent, as evidenced by the turbidity of the culture in MSM supplemented with 25 µg mL^{-1} buprofezin. The viable cell number was determined after plating an aliquot the culture medium. At 0 h, the cell density in the culture medium supplemented with 25 µg mL^{-1} buprofezin was 7.36×10^3 CFU mL^{-1}. Bacterial cell number increased gradually with increasing incubation time. Thus, after 6 and 24 h, bacterial cell number in the culture medium increased to 17.6×10^3 and 19.2×10^3 CFU mL^{-1}, respectively. Within 24 h of incubation, optimum cell density was achieved in MSM supplemented with 25 µg mL^{-1} buprofezin.

Table 10.1 Biochemical characteristics of the bacterium isolated from soil

S. No.	Biochemical test	Result
1.	Starch hydrolysis test	Positive
2.	Gelatin liquefaction test	Positive
3.	Catalase test	Positive
4.	Carbohydrate fermentation test – Glucose	Acid +, gas +
	Sucrose	Acid +, gas −
	Lactose	Acid −, gas −
5.	Indole production test	Positive
6.	Methyl red test	Positive
7.	Voges-Proskauer test	Negative
8.	Citrate utilization test	Negative
9.	Urease test	Positive
10.	Nitrate reduction test	Positive

Thus, buprofezin, a thiadizine insecticide, was utilized by the isolated soil bacterium as a carbon and/or energy source within 24-h incubation, as evidenced by an increase in growth up to 161%.

It has been very well established by Funayama et al. (1986) that under the flooded conditions, [^{14}C]buprofezin gradually decomposed in the soil. After 150 days, 42% of the applied buprofezin remained unchanged. From the first-order decreasing line, the half life of buprofezin in the flooded soil was estimated at 104 days. Correspondingly, the $^{14}CO_2$ trapped in 2 M NaOH increased with time and finally accounted for 24% of the applied radioactivity after 150 days. It cannot be overemphasized that the sterilization of the flooded soil retarded buprofezin decomposition (the extrapolated half life, 432 days) and completely inhibited the evolution of $^{14}CO_2$. Even after 150 days, 80% of [^{14}C]buprofezin was recovered from the flooded soil and the $^{14}CO_2$ trapped in 2 M NaOH was negligible. Furthermore, [^{14}C]buprofezin also decreased gradually in the soil under the upland condition with a half life of 80 days. The formation of $^{14}CO_2$ was observed to be similar to that under the flooded conditions, accounting for 23% of the applied radioactivity after 150 days. Here again, soil sterilization retarded buprofezin degradation (half life 673 days) and completely inhibited the evolution of $^{14}CO_2$ under the upland condition. Thus, these results also reveal a great deal about the microbial utilization of buprofezin. The adaptation of microbial populations most commonly occurs by induction of enzymes necessary for biodegradation followed by an increase in the population of biodegrading organisms (Leahy and Colwell 1990). Even on MSA supplemented with 25 μg mL^{-1} acephate or buprofezin, there was an appreciable growth of *Pseudomonas* sp. after 48 h of incubation at 37 °C (Fig. 10.1). The results of the present study clearly indicate that the bacterial strain is capable of utilizing acephate as sole source of carbon and energy, and has a great potential for use in bioremediation of acephate-contaminated soils.

One of the most remarkable features of soil microorganisms is that they collectively decompose various xenobiotic compounds and return elements to the mineral

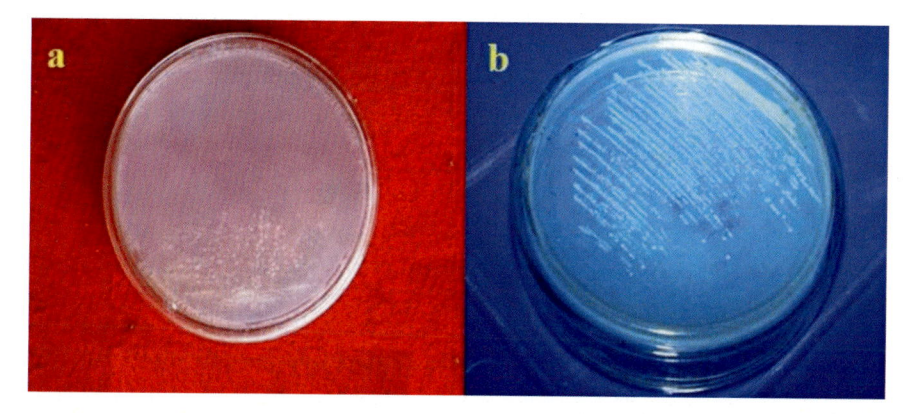

Fig. 10.1 Growth of *Pseudomonas* sp. on MSA supplemented with 25 μg mL^{-1} of (**a**) acephate or (**b**) buprofezin

state for their utilization by plants. They also play important roles in the dissipation of xenobiotic pesticides in the soil. In recent years, the role of soil microorganisms in affecting the persistence of agricultural pesticides has been the subject of two areas of study: to investigate the capacity of microorganisms for rapid degradation of highly persistent or toxic chemicals, and to understand the reduced pesticide efficacy particularly when applied repeatedly under a continuous cropping program. Once present in soil, pesticides may influence the growth and activity of microorganisms in soil essential for maintaining the soil fertility (Wainwright, 1978; Somerville and Greaves, 1987).

Several species of bacteria have been isolated and characterized that can degrade pesticides such as diazinon, monocrotophos, malathion, dimethoate, glyphosate, conmaphos, fenitrothion in liquid medium and soils. Virtually, there are no studies available in the literature on biodegradation of acephate by pure cultures of microorganisms. However, more recently Chai et al. (2010) reported that acephate degraded faster in air-dry soil (t½ 9–11 days) and in soil with field capacity (t½ 10–16 days) than in the wet soils (t½ 32–77 days). The activation energy of degradation was in the range 17–28 kJ mol^{-1} and significantly higher for the soil with higher pH and lower clay and iron oxide contents. Soil sterilization caused a 3- to 10-fold decrease in degradation rates compared to non-sterile soils (t½ 53–116 days) demonstrating that acephate degradation is mainly governed by microbial processes. Likewise, several OP insecticides have been reported to induce the growth of other microorganisms in pure cultures. Megharaj et al. (1986) reported enhanced growth of two cyanobacteria, *Synechococcus elongatus* and *Phormidium tenue,* at all the concentrations of monocrotophos and quinalphos tested. Application of fenamiphos at 0.5–2 kg ha^{-1} resulted in almost doubling of the algal population in soil after 10-day incubation, although the population returned to that of untreated soil by 20-day incubation (Megharaj et al. 1999). In another study, Megharaj et al. (1987) investigated the effects of cypermethrin on *Nostoc linckia*. At 10–50 µg mL^{-1} cypermethrin, growth of *N. linckia* was greatly enhanced. Subramanian et al. (1994) found maximal growth of *Aulosira fertilissima* ARM68 at 50 µg mL^{-1} monocrotophos. Ramakrishnan et al. (2010) opined that most organic pollutants are based on aliphatic, alicyclic or aromatic structures. In the environment, majority of organic carbon available to microorganisms are photosynthetically-fixed carbon compounds. Hence, many of the manmade organic chemicals that have structural similarity to naturally-occurring organic carbon can be easily degraded. However, the manmade chemicals may change the carrying capacity of the environment (i.e., the maximum level of microbial activity that can be expected under a particular environmental condition).

In catabolic processes, a parent pesticide or a primary metabolite is degraded with a concomitant benefit to the microorganism, because it utilizes the compound as a carbon/energy source or nutrient. This occurs with diazinon and parathion, which are hydrolyzed by *Flavobacterium* sp. This bacterium used the hydrolysis products from diazinon and parathion as sole carbon sources (Sethunathan and Yashida 1973). One of the enzymes evolved in this process (parathion hydrolase) has been isolated from *Pseudomonas diminuta* (Mulbry et al. 1986). It has been noted that repeated exposure of soil or aquatic microorganisms to the same OP pesticides results in the proliferation of pesticide-degrading microorganisms, mainly bacteria that have enhanced biodegradation capacity. Some OP compounds such as diazinon, conmaphos, fensulfothion, and chlorfenvinphos were known to be susceptible to the phenomenon (Sethunathan and Pathak 1972; Racke and Coats 1990). Results from several studies (Karpouzas and Walker 2000; Karpouzas et al. 2000, 2004) demonstrated the capacity of some soil bacteria, *Flavobacterium* sp., *Pseudomonas* sp. and *Sphingomonas* sp., to breakdown OP pesticides such parathion, cadusafos, and ethoprophos. Pakala et al. (2007) isolated a bacterium capable of utilizing methyl parathion as a sole carbon and energy source. Xu et al. (2007) reported that a bacterial strain of *Serratia* sp. transformed chlorpyrifos to 3,5,6-trichlor-2-pyridinol (TCP) and that a fungal strain *Trichosporon* sp. was capable of mineralizing TCP. These microorganisms were isolated from an activated sludge. It was observed that the cultures completely mineralized 50 μg mL^{-1} chlorpyrifos within 18 h at 30 °C and pH 8, using a total inoculum of 0.15 g L^{-1} biomass. Thus, such findings are important as bacteria capable of degrading insecticides are potentially useful in bioremediation programs being followed world-wide.

Growth of *Pseudomonas* sp. with Different Concentrations of Acephate and Buprofezin

Growth response of the soil bacterium in MSM supplemented with different rates of insecticides was also investigated. Higher bacterial growth has been observed in MSM supplemented with 50 μg mL^{-1} of either acephate or buprofezin, separately (Fig. 10.2). Contrary to this, bacterial growth was adversely affected at higher concentrations (100, 200 and 500 μg mL^{-1}) of the insecticides. In fact, no growth was seen at 500 μg mL^{-1} of acephate or 200 and 500 μg mL^{-1} of buprofezin. Comparatively, more growth was seen in MSM containing acephate than that was supplemented buprofezin. The data support the rapid utilization of acephate by the bacterial isolate, and have implications for the development of a bioremediation strategy. Such results were found in many studies with different insecticides. Bhaskar et al. (1994) studied the nontarget effects of monocrotophos and quinalphos on culture yield, photosynthetic pigments, and cell constituents in *Anabaena torulosa* at concentrations ranging from 5 to 100 μg mL^{-1}. Higher concentrations (50 and 100 μg mL^{-1}) of monocrotophos and quinalphos unfavorably affected the parameters studied. Moreover, while increasing concentrations of insecticides, nitrogenase activity of the culture was significantly inhibited. Megharaj et al. (1986) also reported that monocrotophos was toxic to *Phormidium tenue* at 50 and 100 μg mL^{-1} and to *Nostoc linckia* only at 100 μg mL^{-1}. Similarly, Obulakondaiah et al. (1993) demonstrated that the growth and metabolic activities of ecologically beneficial nontarget organisms of soil such as *Anabaena torulosa* were greatly altered under the influence of carbaryl, at field application rates, and its hydrolysis product, 1-naphthol, alone or in combination. In contrast, soil application of cypermethrin and fenvalerate enhanced the rate of ammonification and nitrification and that the increased nitrogen-fixing capacity of *Azospirillum* sp., isolated from soils treated with cypermethrin and fenvalerate, lasted for longer periods (Rangaswamy and Venkateswarlu 1993).

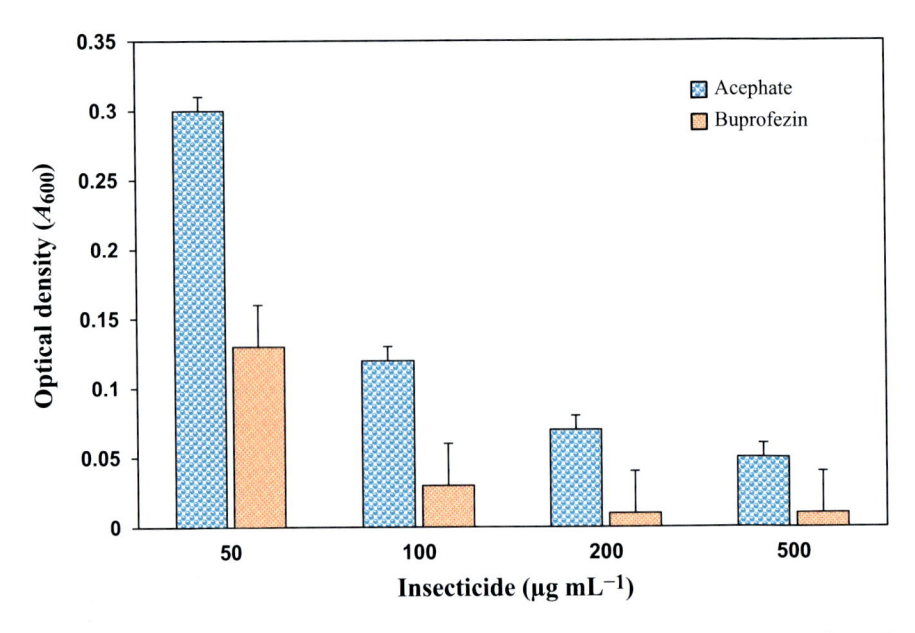

Fig. 10.2 Growth of *Pseudomonas* sp. in MSM supplemented with different concentrations of acephate or buprofezin. Error bars represent standard deviations (n = 3)

Similarly, Megharaj et al. (1990) reported synergistic interaction response with carbaryl and 1-naphthol combinations at concentrations ranging from 10 and 20 μg mL^{-1} for nitrogen fixation in *Nostoc linckia*.

Growth of *Pseudomonas* sp. with Different Carbon and Nitrogen Sources in Presence of Acephate and Buprofezin

In a separate experiment, growth response of the soil bacterium has been studied in MSM containing 50 μg mL^{-1} of either acephate or buprofezin plus 0.01% of different carbon sources viz., glucose, yeast extract and sodium acetate. The organism has shown higher growth in the presence of sodium acetate than glucose or yeast extract (Fig. 10.3). By comparison, growth was more in MSM containing buprofezin than acephate. Thus, carbon sources other than target chemicals may influence the growth rates. Also, growth response of the bacterial strain was assessed with different ammoniacal nitrogen sources, viz., ammonium chloride, ammonium nitrate and ammonium phosphate in MSM containing 0.01% sodium acetate plus 50 μg mL^{-1} of either acephate or buprofezin. Growth of the bacterium was more in MSM containing ammonium chloride than that contained the other two nitrogen sources (Fig. 10.4). There was no growth in MSM containing 50 μg mL^{-1} of buprofezin in the presence of either ammonium nitrate or ammonium phosphate. The effect of

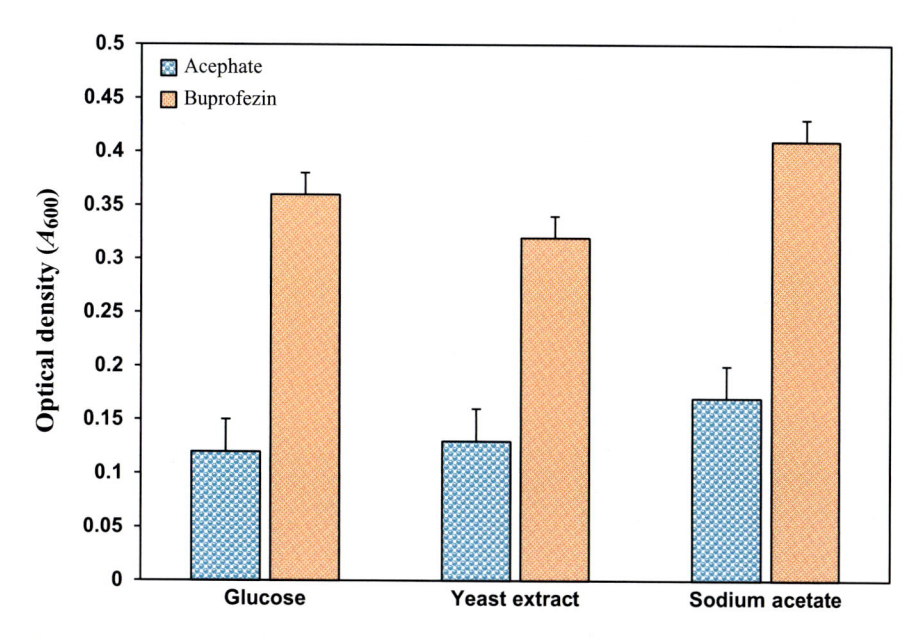

Fig. 10.3 Growth of *Pseudomonas* sp. in MSM supplemented with 0.01% of different carbon sources and an insecticide (50 µg mL^{-1}). Error bars represent standard deviations (n = 3)

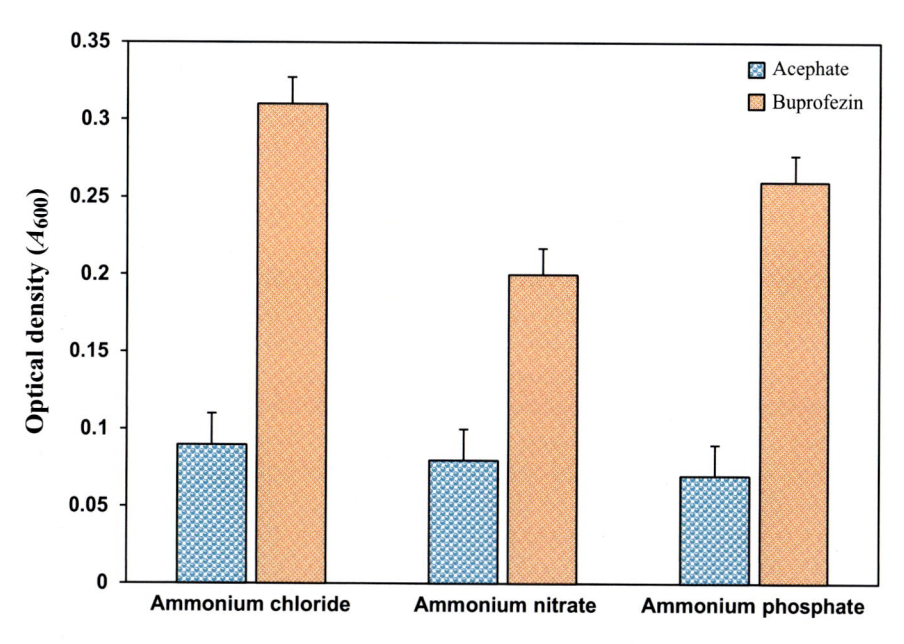

Fig. 10.4 Growth of *Pseudomonas* sp. in MSM supplemented with different ammoniacal nitrogen sources, 0.01% of sodium acetate and an insecticide (50 µg mL^{-1}). Error bars represent standard deviations (n = 3)

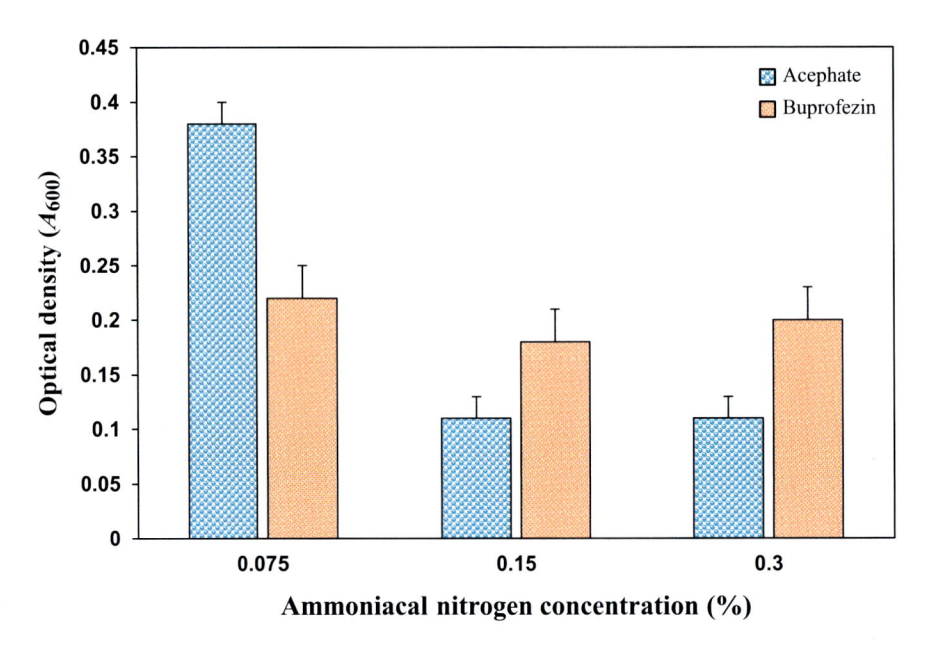

Fig. 10.5 Growth of *Pseudomonas* sp. in MSM supplemented with different concentrations of ammonium chloride, 0.01% sodium acetate and an insecticide (50 μg mL⁻¹). Error bars represent standard deviations (n = 3)

0.01% sodium acetate and different rates of ammonium chloride on growth of the bacterial strain was studied in MSM supplemented with 50 μg mL⁻¹ of either acephate or buprofezin. Indeed, growth of the bacterium was more when ammonium chloride was included in MSM at a concentration which was 50% less than that used in the above experiment (Fig. 10.5). Thus, it is necessary to study and understand the effect of pesticides on soil microflora and the role of pesticide-utilizing capability of microorganisms.

Perrin-Ganier et al. (2001) observed that the addition of nitrogen and phosphorus inhibited the decomposition rate of isoproturon in soil. More recently, Huerta and Vazquez (2010) reported that coffee bean was an adequate nutrient source for bacterial growth and it significantly enhanced DDT and endosulfan biodegradation in comparison with glucose and peptone. Addition of glucose to the culture medium increased bacterial mass of *Alcaligens denitrificans* (from 0.78 mg to 8.4 mg), but DDT metabolism was only 3% increased (Ahuja and Kumar 2003). In contrast, the presence of favorable substrates such as sodium succinate, sodium acetate, sodium citrate, glucose or sucrose inhibited DDT biodegradation by *Serratia marcensces* (Bidlam and Manonmani 2002). Also, sodium acetate and sodium succinate have been reported to inhibit endosulfan degradation (Awasthi et al. 2000). Likewise, a convenient nutrient source should promote growth of microorganisms and pollutant degradation. The present investigation clearly demonstrates the enrichment of bacteria in soil capable of utilizing the two commonly used insecticides, acephate and buprofezin.

References

Adebayo TA, Ojo OA, Olaniran OA (2007) Effect of two insecticides Karate and Thiodan on population dynamics of four different soil microorganisms. Res J Biol Sci 2:557–560

Agnihotri NP, Pandey SY, Jain HK, Srivastava KP (1981) Persistence, leaching and movement of chlorfenviphos, chlorpyrifos, disulfothion, fensulfothion, monocrotophos and tetrachlorvinphos in soil. Indian J Agric Chem 14:27–31

Ahuja R, Kumar A (2003) Metabolism of DDT [1,1,1-trichloro-2,2-bis (4-chlorophenyl) ethane] by *Alcaligenes denitrificans* ITRC-4 under aerobic and anaerobic conditions. Curr Microbiol 46:65–69

Awasthi N, Ahyra R, Kumar A (2000) Factors influencing the degradation of soil applied endosulfan isomers. Soil Biol Biochem 32:1697–1705

Bhaskar M, Sreenivasulu C, Venkateswarlu K (1994) Nontarget effects of monocrotophos and quinalphos towards *Anabaena torulosa* isolated from rice soil. Micribiol Res 149:395–400

Bidlam R, Manonmani HK (2002) Aerobic degradation of dichlorodiphenyl-trichloroethane (DDT) by *Serratia marcescens* DT-1P. Process Biochem 38:49–56

Bunemann EK, Schwenke GD, Van Zwieten L (2006) Impact of agricultural inputs on soil organisms – a review. Aust J Soil Res 44:379–406

Chai L, Wong M, Mohd-Tahir N, Hansen HCB (2010) Degradation and mineralization kinetics of acephate in humid tropic soils of Malaysia. Chemosphere 79:434–440

Chapalamadugu S, Chaudhry GR (1992) Microbiological and biotechnological aspects of metabolism of carbamates and organophosphates. Crit Rev Biotechnol 12:3357–3389

Chapman RA, Harris CR (1982) Persistence of isofenphos and isazophos in a mineral and an organic soil. J Environ Sci Health B17:355–361

Chaudhry GR, Wheeler WB (1988) Biodegradation of carbamates. Water Sci Technol 20:89–94

Chevron Chemical Co. – Ortho Division (1972a) Orthene soil metabolism – Laboratory studies (aerobic). California Department of Pesticide Regulation (CDPR). 108–163:54150. http://www.cdpr.cagov/docs/emon/pubs/fatememo/acephate.pdf

Chevron Chemical Co. – Ortho Division (1972b). Orthene soil metabolism – Laboratory studies (anaerobic). California Department of Pesticide Regulation (CDPR). 108–163:54151. http://www.cdpr.cagov/docs/emon/pubs/fatememo/acephate.pdf

Chevron Chemical Co. – Ortho Division (1972c) Comparison of orthene soil metabolism under aerobic and anaerobic conditions (aerobic). California Department of Pesticide Regulation (CDPR). 108–163:54153. http://www.cdpr.cagov/docs/emon-/pubs/fatememo/acephate.pdf

Chevron Chemical Co. – Ortho Division (1972d) Comparison of orthene soil metabolism under aerobic and anaerobic conditions (anaerobic). California Department of Pesticide Regulation (CDPR). 108–163:54154. http://www.cdpr.cagov/docs/emon/-pubs/fatememo/acephate.pdf

Digrak M (1994) Widely used pesticides in Elazig region of *Bacillus* sp., *Pseudomonas* sp. fragmentation of mixed cultures and by soil microorganisms. PhD Thesis, Firat University Institute of Science and Technology, Elzig

Digrak M, Kazanici F (2001) Effect of some organophosphorus insecticides on soil microorganisms. Turkish J Biol 25:51–58

Digrak M, Ozcelik S (1998) Effect of some pesticides on soil microorganisms. Bull Environ Contam Toxicol 60:916–922

Digrak M, Ozcelik S, Celik S (1995) Degradation of ehthion and methidation by some microorganisms. 35th IUPAC Congress, Istanbul, 14–19 Aug, pp 84

Funayama S, Uchida M, Kanno H, Tsuchiya K (1986) Degradation of buprofezin in flooded and upland soils under laboratory conditions. J Pestic Sci 11:605–610

Gauger WK, Donald JMM, Adiran NR, Matthees DP, Walgenbach DD (1986) Characterization of a streptomycete growing on organophosphate and carbamates insecticides. Arch Environ Contam Toxicol 15:137–141

Ghisalba O, Kuenzi M, Tombo GM, Schar HP (1987) Organophosphorus microbial degradation and utilization of selected organophosphorus compounds: strategies and applications. Chemia 41:206–210

Gorder GW, Dahm PA, Tollefson JJ (1982) Carbofuran persistence in cornfield soils. J Econ Entomol 75:637–642

Gundi VAKB, Reddy BR (2006) Degradation of monocrotophos in soils. Chemosphere 62:396–403

Harkness MR, Mc Dermott JB, Abramowicz DA, Salvo JJ, Flanagan WP, Stephens ML, Mondello FJ, May RJ, Lobos JH, Carroll KM (1993) In situ stimulation of aerobic PCB biodegradation in Hudson River sediments. Science 259:503–507

Huerta BEB, Vazquez RR (2010) Green bean coffee as nutrient source for pesticide degrading-bacteria. In: Mendez-Vilas A (ed), Current research, technology and education topics in applied microbiology and microbial biotechnology. Formatex Research Center, Badajoz. ISBN: (13): 978-84-614-6195-0, 2:1322–1327

Iranzo M, Sain-Pardo I, Boluda R, Sanchez J, Mormeneo S (2001) The use of microorganisms in environmental remediation. Ann Microbiol 51:135–143

Karpouzas D, Karanasios E, Menkissoosh U (2004) Enhanced microbial degradation of cadusafos in soil from potato monoculture: demonstration and characterization. Chemosphere 56:549–559

Karpouzas D, Morgan J, Walker A (2000) Isolation and characterization of ethoprophos-degrading bacteria. FEMS Microbiol Ecol 33:209–218

Karpouzas D, Walker A (2000) Factors influencing the ability of *Pseudomonas putida* strains epl and II to degrade the organophosphate ethoprophos. J Appl Microbiol 89:40–48

Landis WG, Frank JJD (1991) In: Kamely D, Chakrabarty A, Omenn GS (eds) Biotechnology and biodegradation, Methods and applications in biodegradation. Gulf Publishing, Houston, pp 183–201

Leahy JG, Colwell RR (1990) Microbial degradation of hydrocarbons in the environment. Microbiol Rev 54:305–315

Matsumura F (1988) Degradation of pesticides in the environment by microorganisms and sunlight. In: Matsumura F, Murti CRK (eds) Biodegradation of pesticides. Academic Press, New York, pp 67–87

Megharaj M, Prabhakara Rao A, Rao AS, Venkateswarlu K (1990) Interaction effects of carbaryl and its hydrolysis product, 1-naphthol, towards three isolates of microalgae from soil. Agric Ecosyst Environ 31:293–300

Megharaj M, Singleton I, Kookana R, Naidu R (1999) Persistence and effects of fenamiphos to native algal populations and enzyme activities in soil. Soil Biol Biochem 31:1549–1553

Megharaj M, Venkateswarlu K, Rao AS (1986) Growth response of four species of soil algae to monocrotophos and quinalphos. Environ Pollut 42:15–22

Megharaj M, Venkateswarlu K, Rao AS (1987) Influence of cypermethrin and fenvalerate on a green alga and three cyanobacteria isolated from soil. Ecotoxicol Environ Saf 14:142–147

Mulbry WW, Karns J, Keraney PC, Nelson JO, McDaniel CS, Wild JR (1986) Identification of plasmid-borne parathion hydrolase gene from *Flavobacterium* sp. by southern hybridization with *opd* from *Pseudomonas diminuta*. Appl Environ Microbiol 51:926–930

Obulakondaiah M, Sreenivasulu C, Venkateswarlu K (1993) Nontarget effects of carbaryl and its hydrolysis product, 1-naphthol, towards *Anabaena torulosa*. Biochem Mol Bio Int 29:703–710

Pakala SB, Gorla P, Pinjari AB, Krovidi RK, Baru R, Yanamandra M, Merrick M, Siddavattam D (2007) Biodegradation of methyl parathion and *p*-nitrophenol: evidence for the presence of a *p*-nitrophenol 2-hydroxylase in a Gram negative *Serratia* sp. strain DS001. Appl Microbiol Biotechnol 73:1452–1462

Perrin-Ganier C, Schiavon F, Morel JL, Schiavon M (2001) Effect of sludge amendment or nutrient in the biodegradation of the herbicide isoproturon in soil. Chemosphere 44:887–892

Racke KD, Coats JR (1988) Comparative degradation of organophosphorus insecticides in soil: specificity of enhanced microbial degradation. J Agric Food Chem 36:193–199

Racke KD, Coats JR (1990) Enhanced biodegradation of insecticides in Midwestern corn soils. In: Racke KD, Coats JR (eds), Enhanced biodegradation of pesticides in the environment. American Chemical Society Symposium Series 334, American Chemical Society, Washington DC, pp 68–82

Ramakrishnan B, Megharaj M, Venkateswarlu K, Naidu R, Sethunathan N (2010) The impacts of environmental pollutants on microalgae and cyanobacteria. Crit Rev Environ Sci Technol 40:699–821

Ramakrishnan B, Megharaj M, Venkateswarlu K, Sethunathan N, Naidu R (2011) Mixtures of environmental pollutants: effects of microorganisms and their activities in soils. Rev Environ Contam Toxicol 211:63–120

Rangaswamy V, Venkateswarlu K (1993) Ammonification and nitrification in soils, and nitrogen fixation by *Azospirillum* sp. as influenced by cypermethrin and fenvalerate. Agric Ecosyst Environ 45:311–317

Roberts SJ, Walker A, Parekh NR, Welch SJ, Waddington MJ (1993) Studies on a mixed bacterial culture which degrade the herbicide linuron. Pestic Sci 39:71–78

Robertson LN, Chandler KJ, Stickley BDA, Cocco RF, Ahmetagic M (1998) Enhanced microbial degradation implicated in rapid loss of chloropyrifos from the controlled-release formulation suSCon® Blue in soil. Corp Prot 17:29–33

Sethunathan N, Pathak MD (1972) Increased biological hydrolysis of diazinon after repeated application of rice paddies. J Agric Food Chem 20:586–589

Sethunathan N, Yoshida T (1973) A *Flavobacterium* sp. that degrades diazinon and parathion as sole carbon source. Can J Microbiol 19:873–875

Sharmila M, Ramanand K, Sethunathan N (1989) Effect of yeast extract on the degradation of organophosphorus insecticides by soil enrichment and bacterial cultures. Can J Micribol 35:1105–1110

Siddaramappa R, Rajaram KP, Sethunathan N (1973) Degradation of parathion by bacteria isolated from flooded soil. Appl Microbiol 26:846–849

Singh BK, Kuhad RC, Singh A, Lal R, Tripathi KK (1999) Biochemical and molecular basis of pesticide degradation by microorganisms. Crit Rev Biotechnol 19:197–225

Singh BK, Walker A, Morgan JA, Wright DJ (2003) Role of soil pH in the development of enhanced biodegradation of fenamiphos. Appl Envrion Microbiol 69:7035–7043

Somerville L, Greaves MP (1987) Pesticide effects on soil microflora. Taylor and Francis, London

Subramanian G, Sekhar S, Sampoornam S (1994) Biodegradation and utilization of organophosphorus pesticides by cyanobacteria. Int Biodeterior Biodegrad 33:129–143

US EPA (United States Environmental Protection Agency) (1987) Surface water monitoring: a framework for change. U.S. Environmental Protection Agency, Office of Water, Office of Policy Planning and Evaluation, Washington, DC

Visalakshi A, Mohammed AB, Rema DL, Mohandas N (1980) The effect of carbofuran on the rhizosphere microflora of rice. J Microbiol 20:147–148

Wainwright M (1978) A review of the effect of pesticides on microbial activity in soils. J Soil Sci 29:287–289

Xu G, Li Y, Zheng W, Peng X, Li W, Yan Y (2007) Mineralization of chlorpyrifos by co-culture of *Serratia* and *Trhichosporon* spp. Biotechnol Lett 29:1469–1473

Zhi XK, Zhi XP, Sheng CJ, Hu TS, Bao ZF, Hai YH, Xu H (2008) Isolation and identification acephate-degradation bacteria XP-3 and studies on its physiological characterization. Guangdong Agricultural Sciences, Guangzhou. http://en.cnki.com.cn/Article_en/CJFDTotalGDNY200808030.htm

Index